Citizen's Guide to Terrorism Preparedness

Richard Stilp
Armando Bevelacqua

Australia Canada Mexico Singapore Spain United Kingdom United States

Citizen's Guide to Terrorism Preparedness

Richard Stilp and Armando Bevelacqua

Business Unit Director:
Alar Elken

Executive Editor:
Sandy Clark

Acquisitions Editor:
Mark Huth

Development:
Dawn Daugherty

Executive Marketing Manager:
Maura Theriault

Channel Manager:
Mary Johnson

Marketing Coordinator:
Erin Coffin

Executive Production Manager:
Mary Ellen Black

Production Manager:
Larry Main

Production Editor:
Elizabeth Hough

Cover Design:
Julie Lynn Moscheo

Printed in Canada
1 2 3 4 5 XX 06 05 04 03 02

For more information contact
Delmar Learning
Executive Woods
5 Maxwell Drive, PO Box 8007,
Clifton Park, NY 12065-8007
Or find us on the World Wide Web at
http://www.delmar.com

ISBN: 1-4018-1474-3

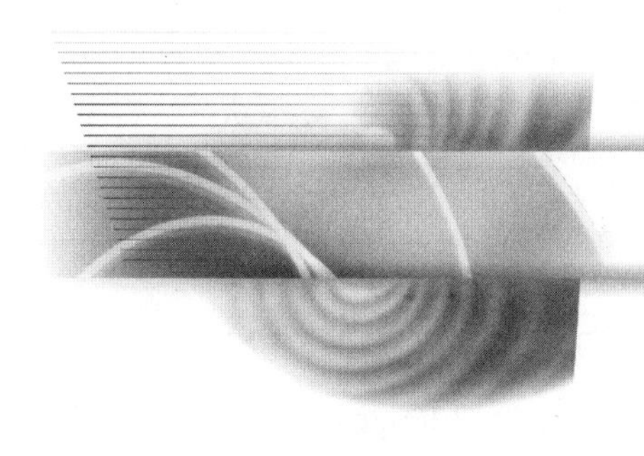

CONTENTS

FOREWORD

Have you ever considered why choices are made and the motivating factors involved? Why did you pick this book up and begin to read it? Perhaps you, as I, fully realize that bioterrorism is no longer science fiction or a historical subject. Nor is it optional reading for those interested in keeping others as well as themselves safe. We cannot afford to read about this topic as an intriguing, tantalizing "what if"? Bioterrorism is a very real threat each and every day. To mitigate this threat with any degree of calm and control, we must be educated and prepared until we feel comfortable handling the threat. Join me, as we prepare for an uncertain future.

—Cynthia Vlasich, RN, Executive Director of Nursing Spectrum

The World Trade Center and Oklahoma City Federal Building bombings demonstrated that terrorism can happen "in our own backyard" and affect our everyday lives. Although dramatic, these events did not affect the lives of every American. Unfortunately, they were viewed as distant occurrences by many. These events were followed by attacks on Americans at the embassies in Nairobi, Kenya, and Tanzania and on the USS Cole. Still, many Americans could not believe that terrorism had taken a hold in today's culture or that the threat of a serious attack on American soil was growing. On September 11th, 2001, Americans were compelled to believe this threat.

Weeks after September 11th, letters containing anthrax reached media and government offices. Some victims died and many were treated for exposure. Concern of an extensive attack using a biological weapon spread across the country. Many questions went unanswered. How could this happen? Will it happen again? Who was behind these acts? Definitive answers were not available. The media presented information to the public but opinions differed from one expert to another and depended on what newspaper was read or news program was watched.

Many education and training efforts have been undertaken to prepare law enforcement officers, firefighters, healthcare providers, and the military to prevent, plan for, and manage terrorist acts. Although these efforts have been extensive, one important group is missing from the preparedness efforts—everyday citizens. They have not received adequate education and training about bombings, or biological and chemical weapons. In fact, most of the knowledge possessed by the average citizen has come from the news media that sensationalize and, in some cases, stretch the facts to provide entertainment and boost ratings.

The intention of this book is to answer the question that has been asked over and over again by the public—"What can I do"? This book brings hard facts and reliable information to the American people who want to be educated and knowledgeable about terrorist acts and how these events can affect their lives.

PREFACE

Times have changed. Hardly a decade ago, the word *terrorism* was rarely heard in the United States. Terrorism was a term used to describe violent acts in other countries. Some consider the turning point of terrorism in the United States to be the World Trade Center bombing in 1993. Now, hardly a day goes by that daily news shows and newspapers are not saturated with stories of possible threats and new actions taken to combat terrorism in the United States.

Starting in the mid-1990s, the federal government has provided training and equipment grants for emergency response agencies. This training and equipment aims to prepare local responders and emergency management agencies in the case of a terrorist act. These much-needed resources help communities that could not prepare on their own. Today, certain acts of terrorism would be handled by emergency response agencies in a much more efficient and coordinated manner. However, regardless of how prepared law enforcement, fire departments, emergency management, or public health officials are, the ability to manage the aftermath of acts of terrorism depends on how the general public perceives and reacts to a terrorist act.

It is unfortunate that very few resources are spent on public awareness and education. Without the public's ability to act and react in a reasonable manner, preparedness efforts by emergency responders will only be partially successful. This book is intended to provide the general public information that can be used to learn about and prepare for acts of terrorism.

Terrorism has become part of our language and a feared aspect of our culture. The only way to combat fear is through education and preparedness. Training and education from the federal government for the general public remains unlikely. It is time for us to prepare and be ready for these unthinkable acts on our own. If we understand the possible acts of terrorism and know what to do during and after an attack, emergency response agencies can perform their functions more safely and with greater efficiency. Together, we can all act to reduce the effects and success of terrorism in America.

The authors and the publisher would like to thank John Kimball from the Federal Emergency Management Agency (FEMA) for reviewing the manuscript and making suggestions for improvement. This is a better book because of Mr. Kimball's contributions.

INTRODUCTION

FAILURE TO PREPARE IS PREPARING TO FAIL.
—BENJAMIN FRANKLIN

This book is not written to generate fear but instead, through the spread of knowledge, to give strength to the American public. Information is presented to put knowledge in the hands of American people, who have the right to be educated and informed about the weapons that could be, and have been, used against them. Unfortunately, in recent years we have seen truck bombs and commercial aircraft used to attack major buildings, and anthrax sent in the mail to cause illness and death. It is difficult to predict what may come next because terrorism has been called the "thinking man's game." Terrorism differs from many other crimes as these heinous acts in particular are planned for months or years with the effects carefully calculated and meticulously designed. It is only through knowledge that we all can act reasonably, responsibility, and with due vigilance to keep our families and ourselves safe.

Terrorism is not an act of war but instead an attempt to change people's views or beliefs about an issue. Terrorist acts are a violent means to generate fear in a population and persuade that population to respond in a manner desired by the terrorist. Because today's society is affected by global influences, terrorists want global attention. This is accomplished through efforts to injure or kill large groups of people, thereby gaining media attention and worldwide awareness for their cause.

The thought of being prepared is not an overreaction to some isolated incidents but a means of giving strength to those who could fall victim. Many of us already prepare for natural disasters like hurricanes, tornadoes, floods, and earthquakes. Others, because of our close proximity to industrial chemical plants, prepare for accidental chemical releases while those near nuclear facilities prepare for radiological emergencies. Citizens in countries like Israel, Northern Ireland, and Iran prepare for bombings, chemical, and biological attacks by having protective equipment and plans in place ahead of time to protect themselves, their families, and their homes. In these countries the threats have been made real through years of terrorist activities

We are entering an era of violent acts against citizens in our own country as well as abroad. These acts are perpetrated by both those who live in our country and those outside of it. In either case, the driving force is usually hatred for our way of life, different beliefs, values, or religion. Simple self-protection, planning, and preparedness efforts will help to keep us safe from these violent acts. There are no guarantees that these acts will not affect any one of us. The time has come to put aside any naiveté that may exist and face the fact that terrorism is part of our lives. Through knowledge and preparedness we can lessen the impact these acts have on our country, our culture, and our values.

1

HISTORY

OVERVIEW

The use of chemical and biological weapons has been going on for centuries. In 1346, the Tartar army hurled the corpses of those soldiers who died of the plague over the Kaffa City walls, infecting residents who were defending the city. It is suspected that some of those who left Kaffa may have started the "Black Death" plague pandemic that spread through Europe. This same tactic was repeated by Russian troops against Sweden in 1710. Another interesting case of biological warfare occurred during the French and Indian War of 1754 to 1767 when the Englishman Sir Jeffery Amherst provided Indians that were loyal to the French, with smallpox-laden blankets. The Native Americans sustained epidemic casualties as a result. Even then, the act of using sickness as a tool of war was considered inhumane.

Historically, the methods of choice for terrorist organizations have been bombings or armed attacks. In fact, even with all of the recent media coverage concerning biological and chemical attacks, bombs are still used by terrorists 70 to 80 percent of the time. Although there were a few attempts to use biological or chemical agents years before, it was not until March 20, 1995 when the Aum Shinrikyo used the military nerve agent Sarin in an attack in the Tokyo subway, that the realization of the use of these weapons against innocent citizens became real.

In the 1950s and 1960s Americans reacted to the threat of nuclear warfare and what could occur in our own homeland. The result was a nationwide effort to train citizens and prepare them for what might have resulted if a nuclear weapon was used. The Civil Defense program was extensive and reached into almost every neighborhood, preparing citizens for the worst. Emergency responders were also trained in the use of Geiger counters so they could survey potentially contaminated areas and determine which areas would be safe after an attack. As the threat of nuclear war faded away in the early 1970s, so did the Civil Defense program. All seemed well until the 1990s, when acts of terrorism began to erupt in our nation. These acts were perpetrated by both American citizens and by foreigners. Whether the violent act is home-grown, domestic terrorism or international terrorism created from a foreign force, we must again prepare our nation to protect itself.

THREAT SPECTRUM

There is no doubt that the events of September 11, 2001 changed our lives from that day forward. The acts of that day brought the feeling of war into the homeland of America. Although there had been previous terrorist attacks in the United States, the magnitude of this event left Americans with a feeling of loss not experienced in recent generations. Americans remember April 19, 1995 when a truck bomb, parked in front of the Alfred P. Murrah Federal Building in Oklahoma City exploded and killed 168 innocent people and caused injury to more than 500. Even before that, in 1993, a large truck bomb entered the below-grade parking garage of the World Trade Center and was detonated. Terrorism on a much smaller scale has been in the United States for decades. We have witnessed attempted assassinations, aircraft hijackings, and bombings on a much smaller scale but it seems that with each subsequent act it appears that the stakes become greater. Other countries such as Israel, Iran, and Northern Ireland have experienced numerous acts of terrorism. Their citizens have recognized the need to become prepared for unexpected and unthinkable acts of violence against them and as a result have trained and prepared both their homes and workplaces for such events.

Although the most recent acts of terrorism in the U.S. have caused fear, outrage, and sadness, some good consequences have also resulted. Patriotism, solidarity, and brotherhood are some of these consequences. American citizens have also finally recognized the need to become prepared for acts of violence perpetrated against them. Although we have witnessed how firefighters, police, and the military forces step up and are willingly to risk losing their lives to protect us, we must also take some responsibility for our own lives. Regardless of how strong our buildings are built, how mighty our armed forces are, or how wise our politicians appear, it is still our responsibility to be aware, take action, and react when danger arises. Failure to do so may mean the difference between life and death.

So what should we be prepared for? The federal government still believes that bombings and armed attacks are the most likely acts of terrorism in our future but other unconventional weapons are becoming more popular and are capable of creating greater fear. Interest in the use of biological, chemical, and radioactive weapons by terrorist organizations has grown in recent years. In fact, the Department of Justice estimates that the threat spectrum from these unconventional weapons is topped by biological agents. When the potential effects from these weapons are evaluated, the ones most likely to be used are biological or chemical agents and the least likely is a nuclear bomb. When evaluating from a different perspective, the weapon with the highest impact is the nuclear bomb or biological agent. Figure 1.1 demonstrates the threat spectrum concept.

TERRORIST GOALS

When terrorists perpetrate a violent act they may have several motives for the attack. These motives can be evaluated to understand the purpose for the attack. The first and primary motive is the advancement of a cause or goal through a violent act. This advancement is accomplished through the media attention that occurs after a terrorist attack. Unfortunately, the greatest impact is accomplished by creating mass injuries and death tolls. The higher the profile of the event, the more national and international attention is brought to the terrorist cause.

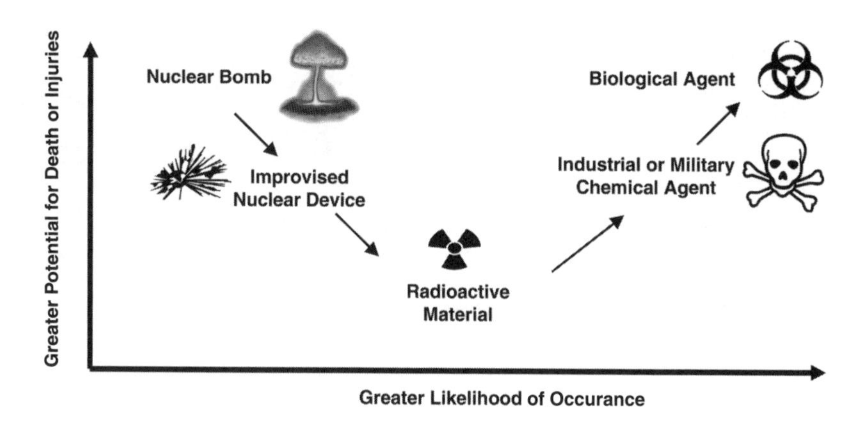

Figure 1–1 *Terrorism Threat Spectrum*

The second motive is to gain more media attention by causing subsequent injuries to first responders on the scene of an attack. Generally speaking, if law enforcement officers or firefighters are injured, the media attention is magnified.

The third motive is to gain attention following a terrorist attack. This bonus attention is accomplished when copycat attacks take place after the original event. After the Columbine High School attack, numerous other school shootings took place. This effectively drew even more attention to the original attack that was carried out because the perpetrators felt that they were being singled out and picked-on by student athletes. Although, not typically thought of as a terrorist attack, the Columbine High School shooting and bombing was a true terrorist attack.

What can be very disconcerting is the fourth motive. It has been referred to as the hidden objective. Some attacks may be used as a test for a larger scale incident or may even be used as a diversion for a different crime. When an incident occurs, especially using a chemical or biological agent or bomb, and no one takes credit for it, there is reason for concern. If an attack does not appear to have a cause or is smaller than anticipated, one should consider that it may be a test for a larger event. The recent unexplained anthrax infections that were deemed unrelated to attacks targeting media professionals or political figures have the markings of a test attack. Figure 1.2 defines the possible objectives of an attack.

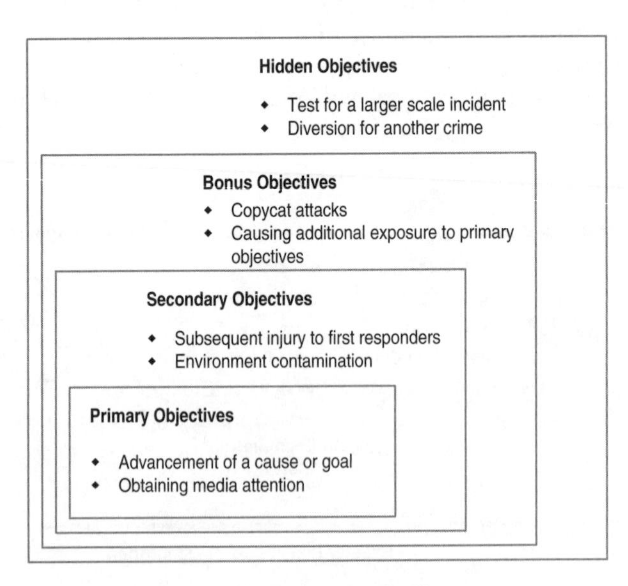

Figure 1–2 *Terrorist Attack Objectives*

PUBLIC INFORMATION

OVERVIEW

Like it or not, media coverage is one reason that terrorism is so successful. The media gives the attention that terrorists need to further their cause. Without the media, terrorist acts would be communicated from person to person and the effects would be very limited, making terrorism largely unsuccessful. This is ironic since one of the things that makes this country great is the ability to freely report any item of importance to the public. In other words, the print and broadcast media have the right to report anything and everything it finds to be newsworthy.

The media can play both a positive or negative role during acts of terrorism. Stories and images involving violence, especially when that violence affects numerous people, commands the attention of the American public. The media therefore plays a powerful role when violent or controversial events occur. American's thoughts and beliefs are shaped by the media who also have the power to alert and educate them. Whether it is television, radio, newspapers, or magazines, the media has the ability to get important information to the public in a timely and efficient manner. During a significant event, the best immediate information comes from the television and radio. If a terrorist event occurs, the first action a person should take is to tune into one of these devices.

RELIABILITY OF INFORMATION

QUESTIONABLE INFORMATION

Generally speaking, the media means no harm and reacts to disasters with an intent to help the situation by providing the best and most reliable information. Unfortunately, with the intent to get the story to an audience first or to add an inventive twist on the story, the resulting story can be filled with unreliable and nonfactual reports. Another news media tactic is to find "experts" with opposing opinions on how an event should be handled. Although the intent of the media in these cases is to provide thought-provoking reporting, this tactic tends to stretch the truth and play the "what if" scenario, causing confusion among those looking for guidance.

RELIABLE INFORMATION

During disastrous situations, emergency services purposely use the news media to get reliable information to the public. Instructions on what to do, which places to avoid, and how to react are all part of the emergency services public information plan. Information coming from law enforcement, fire departments, hospitals, and public health offices are generally the most reliable. Emergency management from municipal, county, and state governments provide instructional news briefings in an effort to direct civilians on needed actions. They may order evacuations, instruct civilians about how to safely stay at home and "protect-in-place," or advise those who may be concerned about an event.

News media in this country are unlike most other countries as they have the guaranteed right to broadcast all aspects of the news. The news media seek to provide an untainted and uncensored representation of what has occurred but their messages are often not without opinion. Be careful what you believe and interpret as fact.

EVALUATING MEDIA REPORTS

These simple rules are helpful in determining the reliability and usefulness of media reports:

- Listen to the news and information bulletins provided by emergency services. This is the most reliable information available to the public.
- Information bulletins and interviews from law enforcement, fire departments, and hospitals are generally factual and not sensationalized.
- State or local offices of emergency management provide the guidance about what to do in a disaster. These instructions will include information about evacuation, safely staying at home, or where and how to seek medical care.

- Sensationalized editorials and the use of controversial experts are entertaining but should not be relied upon for factual information or guidance.
- Do not be drawn into the media "feeding frenzy." Remember to be a critical audience member and stick to the facts.

BIOLOGICAL AGENTS

OVERVIEW

The thought of being infected by a deadly disease or poisoned by a biological toxin is truly a frightful one. It is this fear that may stimulate a terrorist to choose a biological weapon for an attack. Unfortunately, these agents are not difficult to cultivate but instead are surprisingly easy to produce by someone with a very limited knowledge of microbiology. The recent use of anthrax as a weapon sent through the mail has brought international attention to these agents. Although the method used was a very inefficient way to spread the bacteria, fear was generated across the country as buildings were closed, hazardous materials teams responded in full gear, and mail handlers began using gloves and masks to protect themselves from the mail.

Organized research for biological warfare gained momentum during World War II. Prior to this time, biological agents were used to inflict harm on individuals, but the biological agents were, for the most part, not cultivated or produced in mass quantities for a warfare effort. History does reveal that plague-infected corpses were flung over fortress walls in the 14th century and smallpox-ridden blankets were given to the American Indians during the French and Indian War of 1754. True biological warfare research, however, did not take place until much later.

Biological agents are made from a variety of microorganism and biological toxins. While microorganisms are living viruses and bacteria, biological toxins are chemical compounds produced by plants, animals, or microbes that are poisonous to humans. Microorganisms have the ability to establish mortal infections in their victims--some causing diseases that are contagious and spread from one victim to another. Although many of these organisms are recognized as military-type weapons, many others can be cultivated and introduced into the environment with the intention of inflicting harm to a targeted civilian population.

BACTERIA

Bacteria are single-celled microorganisms that are plantlike in structure. They vary in size from about half of a micron to around 10 microns. In other words they are very small and can only be seen under a microscope. Bacterial warfare agents include living cells of anthrax (*Bacillus anthracis*), plague (*Yersinia pestis*), cholera (*Vibrio cholerae*), Q fever, and tularemia (*Francisella tularensis*). These microorganisms are grown in an artificial medium and are then transformed into a weapon-grade substance by concentrating them in solutions or spores. *Bacillus anthracis* and *Clostridium botulinum* (responsible for boutlism) have the ability to spore (becoming seedlike), and live for long spans of time in unfavorable conditions before entering the body and establishing infection. Anthrax spores are particularly hardy and have been known to live in the environment for as long as 40 years. This makes anthrax the ideal biological weapon as it can be produced, loaded into devices for dissemination, and stored for years until its use is needed. The hardiness of the spore is why there is so much concern about decontaminating buildings that have been contaminated with this bacteria.

VIRUSES

Viruses are smaller than most bacteria and live on or within other cells using the host cells' machinery for metabolism and reproduction. The illness experienced from a viral infection is the result of destruction to host cells by the parasitic action of the virus. Viruses cannot be grown in an artificial medium but only in a medium that contains living host cells. Each virus needs a particular type of host cell, making the production of viruses for warfare or terrorist use complicated and expensive.

Virus infections generally have no easy cure. Although there has been some progress in the production of antivirus medications they are only moderately successful with selected viral infections. The difficulty in treating viral infections seems to be the reason that the use of viruses has so eagerly been researched and developed by militaries. The cost and complexity involved in producing viral weapons means it is unlikely that low budget organizations or private individuals will use these biological weapons.

Military forces from many countries have experimented with virus use for warfare weapons. Sections to follow will give an overview of those viruses considered to be the most threatening. These warfare viral agents include Variola virus (smallpox), Venezuelan Equine Encephalitis (VEE), and Viral Hemorrhagic Fever (VHF). The most devastating and concerning one of these agents is smallpox—it is highly contagious, has no cure, and its vaccine is available only in limited quantities.

BIOLOGICAL TOXINS

Biological toxins are toxic substances originating from animals, plants, or microbes. Comparatively speaking, biological toxins are more toxic than most chemicals used and produced in industry. For that matter, some biological toxins are more toxic than any chemicals that man is capable of producing. These toxins are not volatile, making them not as suitable for battlefield use as many of the toxic chemicals are. Instead, these agents are excellent for contaminating food sources, water supplies, and specific targeted individuals. Toxins that have been considered for military use include botulinum toxins (botulism), Staphylococcal Enterotoxin B (SEB), Ricin, and Tricholthecene Mycotoxins (T2).

BACTERIAL AGENTS

ANTHRAX (*BACILLUS ANTHRACIS*)

Anthrax (Ragpickers' disease, Woolsorters' disease) has long been the biological weapon of choice. It was first prepared as a weapon in the United States in the 1950s and continued to be produced until the offensive biological weapon program was terminated. Anthrax is easily grown and can be kept almost indefinitely under the proper conditions. In the spore (dormant) form, *Bacillus anthracis* is easily stored for use when it is needed. There are at least 17 countries that have developed an anthrax weapon program. In 1970, the World Health Organization (WHO) concluded that the release of 50 kilograms or 110 pounds of aerosolized anthrax upwind of a population of five million could lead to an estimated 250,000 casualties and 100,000 deaths, thus substantiating the perceived threat from *Bacillus anthracis*. Because of the efficiency and durability of the *Bacillus anthracis* spore, its use as a biological weapon has brought great concern to the United States military and more recently, to the public. Anthrax spores sent in the mail to various organizations have generated much fear in our country. The fallout from letters that were mailed to political and media offices has resulted in letter handlers becoming ill and dying, furthering the fear suffered by the general public, in addition to the intended targets.

Bacillus anthracis is a spore-forming bacteria that causes a rapidly progressing infection. Once the infection is established, the victim is said to have "anthrax." *Bacillus anthracis* derives its name from the Greek word for coal, "anthrakis" because of the black, coal-like skin lesions it creates. Anthrax can develop from the inhalation, ingestion, or exposure of nonintact skin to *anthracis*. The spores are hardy and can remain viable for more than 40 years.

The Aum Shinriko religious sect spent millions of dollars developing an anthrax weapon and on at least eight occasions released anthrax or botulism. Because of the difficulty encountered in the dispersion, they failed to produce any infection and eventually concentrated their efforts on the nerve agent Sarin. Sarin was eventually used in both Matsumoto and Tokyo, Japan, but was only moderately successful as a lethal agent.

The most devastating experience with inhalation anthrax occurred in 1979 in Sverdlovsk, Russia when a military biological facility accidentally released weaponized spores into the air from a faulty laboratory ventilation system. When the deaths started to occur, the World Health Organization (WHO) began an investigation to find the cause. The WHO was told by the Soviet Ministry of Health that the deaths were due to contaminated meat, but in the summer of 1992, President Boris Yeltsin acknowledged that the deaths were due to an accidental release from a military microbiology facility, Compound 19. Although initially believed to be much higher, a total of 77 anthrax cases were reported including 66 deaths.

The skin (cutaneous) infection is caused when bacteria enters through cuts or breaks in the skin causing a localized infection. Within two to six days after the infection, a depressed black scab develops. Recently, cutaneous infections have resulted from handling letters containing anthrax spores, and the spores then touching open skin. Skin infections not promptly treated can progress into a body-wide (septicemic) infection that has a mortality rate of 5 to 20 percent. Because cutaneous anthrax does not progress as rapidly as the pulmonary or gastrointestinal version of the infection, it is usually not fatal if treatment is started before the onset of septicemia.

Unintentional infections are usually associated with the ingestion of meat from infected animals. It generally takes fewer ingested spores to develop a gastrointestinal infection than it does to develop an inhalation infection. The incubation period is from one to seven days after ingestion and is characterized by abdominal pain, bloody diarrhea, nausea, vomiting blood, and fever. The fatality rate of gastrointestinal infection approaches 50 percent. If not rapidly treated, the gastrointestinal form of anthrax will be fatal as it progresses to sepsis.

The most dangerous and devastating form of the bacterial infection is inhalation (pulmonary) anthrax. It is estimated that the inhalation of 8,000 to 50,000 spores is needed to establish an infection. This may seem like a large quantity but even with these numbers it would take a microscope to visualize them. The spores are very small microparticles that measure between three to five microns in size and are easily breathed into the lungs. Because the spores are microscopic, the number of spores needed to infect a person cannot be seen, leaving the victim unaware that the exposure has taken place. Once inhaled, the spores move through the airways where they are able to reach the very fine airways and the final ends of the airways called the alveoli. Alveoli are the part of the lungs that allow the exchange of oxygen with the blood. From there, the spores are carried out of the lungs by an organism called a microphage to an area between the lungs so the body can attempt to naturally disinfect it. When natural disinfection fails it is this area where an infection is established after the spores are inhaled.

The illness from inhalation takes place in two phases. The initial phase causes flulike symptoms, followed by a significant improvement. These flu symptoms include fever, weakness, and headache. The initial improvement stage is perceived by the victim as a recovery and therefore he or she usually will not seek any further medical care. If a diagnosis of anthrax infection is confirmed or even suspected during the first phase, aggressive treatment utilizing antibiotics is somewhat successful. Unfortunately, because of the rare occurrences of this disease, many of those patients presenting in the first phase will be given supportive treatment, be discharged, and will subsequently die.

The second phase of pulmonary anthrax develops suddenly with severe respiratory symptoms, low blood pressure, and shock. This phase generally lasts fewer than 24 hours and most often ends in death from the release of toxins generated from the developing infection. Antibiotics have no effect on the toxins being generated during this infectious process. Because these toxins are fatal to the victim, the use of antibiotics at this phase usually does not change the outcome. During this phase, the mortality rate approaches 100 percent .

Quick Reference 3.1 summarizes important characteristics of anthrax.

CHOLERA (*VIBRIO CHOLERAE*)

The bacteria responsible for a cholera infection gains entrance into the body through contaminated food and water sources. It is an old disease that spreads rapidly when proper sewage precautions are not taken. Its rapid spread and the ease of contamination are reasons that it has been investigated as a wartime biological agent. In countries where sewage disposal is not carefully monitored, cholera has infected and caused the death of thousands of people.

The bacterium attaches to the tissue of the small intestine, causes an oversecretion of fluid, and overwhelms the large intestine's ability to absorb the fluid. This leads to diarrhea, severe dehydration, and in the worst cases, low fluid volume shock. This condition is similar to that of someone who has a traumatic injury and looses too much blood.

The signs and symptoms begin within 12 to 72 hours of exposure and include vomiting, intestinal cramping, and headache. Five to 10 liters of fluid loss can be expected per day. If not aggressively treated with IV fluids and antibiotics, the fluid loss quickly becomes severe and leads to shock and death.

QUICK REFERENCE 3.1 ANTHRAX (*BACILLUS ANTHRACIS*)

TRANSMISSION
Inhalation of spores
Open skin exposure to spores
Ingesting contaminated food

SIGNS AND SYMPTOMS
Fever, weakness, fatigue, dry cough, and chest pain
Abrupt onset of shortness of breath leading to respiratory failure

DIAGNOSIS
Positive chest x-ray, blood/sputum culture

TREATMENT
Penicillin g, 2 million units IV every 2 hours for 4 weeks and Streptomycin 1 gm IM twice a day for 4 weeks OR
Chloramphenicol 500 mg IV every 6 hours for 4 weeks OR
Ciprofloxacin 400mg IV every 8 hours for 4 weeks

PREVENTION
Ciprofloxacin 500mg by mouth twice a day for 8 weeks
Doxycycline 100mg by mouth twice a day for 8 weeks

VACCINE
One dose at 0–2–4 weeks then at 6-12-18 months with annual booster vaccination for anthrax. Available through the CDC

ISOLATION
None required

Treatment includes fluid replacement using intravenous solutions. Antibiotics will shorten the duration of the infection and will kill the infecting microorganisms. It is difficult to become infected through direct contact with an infected person but precautions should be taken to reduce contact with body fluids. A bleach solution of 1:10 with water should be used as a decontaminating agent for any materials or equipment that has contacted body fluids.

Because our domestic water supplies are treated with chlorinating solutions, it would be difficult to infect this water with cholera so the threat from this bacteria is not as great as others discussed in this chapter.

Quick Reference 3.2 summarizes important characteristics of cholera.

QUICK REFERENCE 3.2 **CHOLERA (*VIBRIO CHOLERAE*)**

TRANSMISSION
Drinking water contaminated with bacteria

SIGNS AND SYMPTOMS
Begin after 12–72 hours post exposure
Vomiting, headache, intestinal cramping with little or no fever followed by painless, voluminous diarrhea

DIAGNOSIS
Microscopic exam of stool samples

TREATMENT
Fluid and electrolyte replacement
Antibiotics including:
- Tetracycline
- Ciprofloxacin
- Erythromycin

PREVENTION
A vaccine is available but provides only about 50% protection that lasts about 6 months
Given at 0–4 weeks with booster every 6 months

ISOLATION
Blood and body fluids precautions

PNEUMONIC AND BUBONIC PLAGUE (*YERSINIA PESTIS*)

The plague has had a colorful world history. In 541 AD, the first great plague began in Egypt and spread across Europe, North Africa, and central and southern Asia. During a period of four years it is estimated to have killed 50 to 60 percent of the population. The second plague pandemic began in 1346 and spread throughout the Middle East, killing one third of the European population and more than 13 million in China. Doing so, it earned the name "Black Death".

Because the disease was so prevalent and so many deaths were associated with the disease, a childhood nursery rhyme describing what occurs as a result of the plague has been recited into modern times. The rhyme states; "Ring around the rosie" (describing the red ring around the infected lymph node), "pocket full of posies" (the smell of death was so overwhelming that people would carry pockets full of fragrant flowers to hold under their noses), "ashes, ashes, we all fall down" (ashes are related to burning the bodies and victims dying). Another lasting impression of the disease is related to a plague victim's coughing and sneezing. Since so many people were dying from the plague and priests could not give last rights to all of the victims, they gave other church members the ability to issue last rights. This was done by saying, "God bless you" after plague victims coughed or sneezed. This has become a custom of well wishing that is practiced even today.

Although much has been done to improve the living conditions that contributed to plague outbreaks, there remains a concern that terrorists, using the bacteria in an attack, could still pose a serious threat. Natural occurrences of plague in the United States happen at an average of 13 annually, with most cases occurring in New Mexico, Arizona, California, and Colorado. The disease is endemic to that region because the bacteria causing the infection is found in a particular rodent population living in those western states.

During the 1950s and 1960s, the United States developed the bacteria *Yersinia pestis* as a biological weapon. The Soviet Union reportedly had more than 10 institutions and thousands of scientists working to develop their plague-based weapon. Japan reportedly used infected fleas against China, dropping them in areas of enemy troops, as a way of spreading the plague. In 1970, the WHO assessed a worst-case scenario of a dissemination of 50 kilograms or 110 pounds of Yersinia pestis in an aerosol cloud over a city of five million. They estimated the results to be 150,000 cases of pneumonic plague of which 36,000 would die.

The plague can have three presentations: bubonic, septicemic, and pneumonic plague. All are the result of an exposure to the bacteria, *Yersinia pestis*. The bubonic form is the fleaborne disease transmitted from an infected rodent. Direct contact of open skin with infected tissue or fluids can also cause a bubonic form of the disease. Once exposure takes place, the bacteria is carried to the lymph nodes where an infection is established. Lymph nodes are part of a system in the body that works to kill organisms as they gain access to cause infection. In the case of bubonic plague, the bacteria is very strong and will successfully cause infection in the lymph nodes. This process typically takes between two to eight days after exposure. Because infected fleas usually bite the legs of a victim, the groin lymph nodes are the most often affected (90%). As the infection develops, the lymph nodes will become very painful, swollen, and hot to the touch before ulcerating. If left untreated, septicemic (full-body) infection results and is characterized by fever, chills, prostration, abdominal pain, shock, and bleeding into the skin and other organs. The mortality rate for this level of infection is 50 to 60 percent. In some cases, septicemia can spread an infection to the lungs causing the pneumonic (lung) form of the disease.

Pneumonic plague causes fever, chills, cough and difficulty breathing, rapid shock, and death. It is spread from person to person by droplet form through talking, coughing, or sneezing. The incubation period for a pneumonic exposure is one to four days and is dependent on the amount of inhaled bacteria and the general health of the victim. The pneumonic form of the disease is characterized by overwhelming pneumonia, fever, bloody sputum, chills, and cough. The death rate for pneumonic plague is over 50 percent and approaches 100 percent if treatment is not instituted within 24 hours of the onset of symptoms.

Any patient diagnosed with bubonic plague and a cough or any person with pneumonic plague is infectious and can transmit the infection through coughing, sneezing, or talking. Anyone in close contact with these patients must wear appropriate protective equipment including a mask and eye protection. Victims should also wear a mask to limit transmission of the disease. Patients are considered contagious until antibiotic therapy has been given for a complete 48 hours. Antibiotic treatment for those exposed to the disease should continue for 7 to 10 days.

Quick Reference 3.3 summarizes important characteristics of plague.

QUICK REFERENCE 3.3 PLAGUE (*YERSINIA PESTIS*)

TRANSMISSION

Pneumonic transmission from droplets (cough, sneezing, talking) or by aerosol dispersion by terrorist
Bubonic from infected flea
Septicemic

SIGNS AND SYMPTOMS

Pneumonic: after 2–3 days presents with pneumonia, fever, chills, bloody sputum, respiratory failure, shock and organ failure
Bubonic: swelling in groin or axillary area at lymph nodes

DIAGNOSIS

X-ray evidence of bronchopneumonia, cultures of blood, sputum, or aspirate of bubo

TREATMENT

Pneumonic plague requires immediate treatment as mortality is very high with delays of more than 24 hours
Streptomycin 2 gm IM every day for 2 weeks OR
Chloramphenicol 2gm IV every day for 2 weeks OR
Doxycycline 200mg initial dose then 100mg IV every 12 hours for 2 weeks

PREVENTION

Doxycycline 100mg by mouth twice a day for 7 days

VACCINE

Available through the CDC

ISOLATION

Droplet precautions for pneumonic plague placed in private room

TULAREMIA (*FRANCISELLA TULARENSIS*)

Tularemia was weaponized by the U.S. during the 1950s but was discontinued with the termination of the offensive biological weapon program. It is reasonable to believe that other countries are still actively continuing a biological program that contains tularemia as one of the weapons.

Tularemia is also known as rabbit fever and deer fly fever because these are both agents for the natural transmission of the disease. Blood and body fluids of an infected person or animal or the bite of an infected deer fly, tick, or mosquito transmits the disease. Inhalation of aerosolized bacteria would cause a typhoidal tularemia infection causing respiratory symptoms in 2 to 10 days. This organism is persistent for weeks in water, soil, or animal hides. Because this bacteria is resistant to freezing, once it is frozen into meat, especially rabbit meat, it can persist for years.

Signs and symptoms of an infection include local ulcerations, swollen lymph nodes, fever, chills, and headache. Typhoidal symptoms include fever, headache, substernal discomfort, and nonproductive cough. Even left untreated, the mortality rate of this infection is about 5 percent.

Treatment includes antibiotic therapy that produces excellent results. Secondary infection is unusual so strict isolation is not needed and only the typical personnel protection against secretion and lesions is required.

Quick Reference 3.4 summarizes important characteristics of tularemia.

QUICK REFERENCE 3.4 TULAREMIA (*FRANCISELLA TULARENSIS*)

TRANSMISSION
Inhalation of aerosol from bioterrorism attack
Open skin contact with infected animal

SIGNS AND SYMPTOMS
Inhalation causes respiratory distress, chest pain, cough, and body wide infection

DIAGNOSIS
Cultures of blood, sputum, urine, and skin lesions

TREATMENT
Streptomycin 1 gm IM every 12 hours for 14 days OR
Gentamicin 3–5mg per kilogram IV every day for 14 days

PREVENTION
Doxycycline 100mg by mouth twice a day for 14 days

VACCINE
Live attenuated vaccine provided through state public health department

ISOLATION
None required

Q FEVER (*COXIELLA BURNETII RICKETTSIA*)

Q Fever (Query Fever) is also one of the weapons previously kept in U.S. arsenals. It naturally occurs as an infection found in sheep, cattle, and goats and is an occupational hazard in industries involving these livestock. Q Fever is very infectious through inhalation of infected aerosolized material. It is reported to cause an infection with the inhalation of as little as one organism. The ease at which this microorganism could be harvested and grown and the probability of infection when inhaled make this an agent that terrorists could use efficiently.

The onset of symptoms begins between 10 to 20 days after exposure and is usually self-limiting. This agent is meant to disable persons and cause panic but does not normally cause death. The illness caused by Q Fever usually lasts between two days to two weeks and is characterized by fever, headache, fatigue, and in some cases pneumonia and chest pain.

Treatment involves antibiotic therapy and supportive care. Even untreated, the illness is incapacitating but rarely causes death. Caregivers need only protect themselves from any contaminated materials or a contaminated environment. Decontamination is accomplished with soap and water.

Quick Reference 3.5 summarizes the important characteristics of Q fever.

QUICK REFERENCE 3.5 Q FEVER (*COXIELLA BURNETII*)

TRANSMISSION
Airborne rickettsia from infected sheep, cattle, and goats
Aerosolized as a biological weapon

SIGNS AND SYMPTOMS
Fever, cough, and chest pain within 10 days after exposure. Illness lasts from 2 days to 2 weeks

DIAGNOSIS
From blood laboratory analysis

TREATMENT
Tetracycline by mouth for 5 to 7 days
Doxycycline by mouth for 5 to 7 days

PROPHYLAXIS
Tetracycline may delay but not prevent onset of symptoms

VACCINE
Vaccine is effective but severe local reactions to this vaccine may be seen in those who already possess immunity

ISOLATION
None required

SALMONELLOSIS (*SALMONELLA TYPHIMURIUM*)

One of the most common types of food poisoning is caused from salmonella. Although this bacteria has not been formally adapted as a military biological weapon, it was used by the religious extremist group, the Bhagwan Shree Rajneesh to contaminate local restaurant salad bars. This was carried out in an effort to affect the outcome of a local election. Their plan was to make enough non-supporters sick so that the majority of those voting would favor their interests and vote for them. Although unsuccessful in their attempt to change the election, some 751 people became ill and 45 were hospitalized. Naturally occurring infections of salmonella are caused from ingesting the organism in food contaminated with infected feces containing the bacteria.

Terrorist use of salmonella to harm a targeted group would be a simple task. Once a food source has been identified as contaminated it would be easy to mix it with noncontaminated food and then distribute it to be ingested by the victims. Since the prevalence of these bacteria among meat products and poultry is high, using meat byproducts to cultivate the bacteria is a reasonable way to obtain and produce enough of the microbe to cause harm to a large number of persons. Spreading the bacteria on foods that are not normally cooked, such as salads, is one way to ensure that the bacteria will not be destroyed prior to eating it.

After ingestion of contaminated food, the symptoms begin within 8 to 48 hours. The victim experiences fever, headache, abdominal pain, and watery diarrhea that may contain blood and mucous.

Infected victims can cause secondary infection if the caretaker does not protect himself or herself from body fluids. The bacteria are killed with heat or bleach solutions so all contaminated items should be washed in a bleach solution with hot water.

Quick Reference 3.6 summarizes important characteristics of salmonella.

VIRAL AGENTS

SMALLPOX (VARIOLA VIRUS)

Smallpox is a disease caused from an exposure of variola virus. Although declared eradicated from the world population in 1980 by the WHO, there remains fear that a state-sponsored terrorist group could use it as a biological weapon. Because of the high fatality rate and transmissibility of the virus, it now represents the most serious terrorist threat involving a biological weapon.

The reason for concern is related to the fact that, although the disease was declared eradicated, the virus still exists in laboratories in at least two locations. One is in the Centers for Disease Control and Prevention in Atlanta and the

QUICK REFERENCE 3.6 SALMONELLAE (*SALMONELLA TYPHIMURIUM*)

TRANSMISSION
Causes one of the most common types of food poisoning

SIGNS AND SYMPTOMS
Symptoms occur in 8–48 hours and include fever, headache, abdominal pain and watery diarrhea

DIAGNOSIS
Laboratory microscopy

TREATMENT
Supportive
Some severe cases may require antibiotic consisting of ampicillin, amoxicillin, or chloramphenicol

PREVENTION
None

VACCINE
None

ISOLATION
None required

second is at Vector in Novizbersk, in the former Soviet Union. It is further suspected that specimens may also exist in other locations including North Korea. North Korea is believed to have acquired it from the Soviet Union.

Smallpox received its name at the end of the fifteenth century in England. Originally called small pokes (poke means sac) it was used to distinguish the illness from syphilis which was then called great pokes. Before vaccinations were available almost everyone contracted the disease. With an aggressive vaccination program the incidence of smallpox was dramatically reduced by the 1970s. The last naturally contracted smallpox death was Ali Maow Maalin, a cook in Somalia, Africa who died in October 1977. Then in 1978, a British medical photographer, Janet Parker died after an accidental exposure at the University of Birmingham, England. There have been no subsequent cases of smallpox since that incident in 1978.

Variola occurs in two principle forms, one causes variola minor, having a fatality rate of about one percent. The other causes the more serious variola major. Both forms share the same incubation period, symptoms, and treatment. The incubation period after exposure averages 12 days. The onset of symptoms begins with generalized weakness, fever, rigid muscles and joints, vomiting, headache, and backache, with some developing delirium. Lesions in the mouth and throat appear early in the illness and release large amounts of virus into the saliva. This is the earliest period of infectivity and occurs before any outward signs of the infection are noted. The smallpox rash appears several days after the other symptoms and progresses from a fine rash to a more defined raised red rash, and finally into pustules or puss filled blisters. This rash usually appears all at once then progresses in severity. The progression of the smallpox rash differs from varicella virus (chicken pox) in several ways. Smallpox begins primarily on the face and extremities and in only small numbers on the trunk. This presentation is the opposite of chicken pox that usually begins on the trunk of the body and spreads to the arms and face and develops gradually over a period of days. The illness remains contagious for about three weeks or until all scabs separate.

Suspicion of the illness should be made based on the symptoms with which a person presents. It is difficult to immediately confirm if the rash is smallpox or some other illness. Electron microscopy can provide a higher suspicion of the virus but the microscopes used are rare and only available at some medical centers or laboratories. Cases thought to be smallpox are required to be reported immediately to public health authorities. In an effort to isolate any outbreak, a strict quarantine with respiratory isolation will probably be ordered for a time frame of at least 17 days and will include all persons in direct contact with the index case.

Smallpox spreads directly from person to person through droplets and through the airborne virus created during coughing, sneezing, or talking. It is more contagious during the pre-eruptive period. It is estimated that 30 percent of those exposed through close contact will develop the disease and 30 percent of those will die five to seven days after the onset of symptoms. Protection from infected victims is accomplished through avoiding all contact with blood, body fluids, clothing, or bedding of a victim. Extensive efforts to protect from respiratory exposure should be taken including using a respirator mask having a rating of N95. These masks are commonly used in health care to protect from tuberculosis. There is no drug available to directly treat smallpox, so care is supportive in nature.

The smallpox vaccine was last given to the general population in the United States through the early 1980s and to some military personnel up until 1989. Although limited immunity remains for those who were given the vaccination, they are now susceptible to the virus. The vaccine is known to have its best protective effects for seven to ten years. A booster is needed to again provide the best protection. Currently there are no plans to once again provide smallpox vaccinations or boosters to the American public, although an outbreak of even one case would certainly stimulate a widespread vaccination effort.

Quick Reference 3.7 summarizes important characteristics of smallpox.

QUICK REFERENCE 3.7 SMALLPOX (VARIOLA VIRUS)

TRANSMISSION
Inhalation of aerosol or contact with skin lesions
Bed linen also infectious

SIGNS AND SYMPTOMS
Weakness, fever, chills, headache, delirium
Rash beginning on hands/arms and legs/feet
Rash progresses to skin lesions
All lesions at the same stage unlike chicken pox
Pneumonia is associated with 50% mortality

DIAGNOSIS
Electron microscopic exam of vesicular or pustular fluids

TREATMENT
Cidofovir IV is investigational
Treatment is supportive

PREVENTION
Vaccine available through CDC

ISOLATION
Strict respiratory isolation
Strict quarantine for at least 17 days from index case

VENEZUELAN EQUINE ENCEPHALITIS (VEE)
A FORM OF VIRAL ENCEPHALITIS

VEE causes a naturally occurring disease in horses, mules, burros, and donkeys. The virus is typically found in South America, Central America, Mexico, Trinidad, and Florida. The disease occurs as a result of a bite from a mosquito carrying blood from an infected animal.

When the disease occurs naturally there is always an outbreak among the equidae (horse, mules, etc.) population before humans are affected. To the contrary, if the virus is spread with the intent to cause disease in humans the outbreak will not initially involve equidae during the early outbreak. Another clue that the outbreak is not of natural origin is the occurrence of the disease outside of its natural geographical area.

VEE was weaponized by the United States during the 1950s but was later destroyed when the offensive biological weapon program was terminated. It is suspected that other countries have experimented with VEE but the extent to which those countries were successful in weaponizing the agent is unknown. It is reasonable to believe that since this virus would be easy to acquire, attempts to cultivate this virus by terrorists present a possible threat.

Once a victim is exposed to the virus, an incubation period of between one and five days is followed by a sudden onset of symptoms. VEE generally establishes an infection in the covering membrane (meninges) of the brain and within the brain itself, so the symptoms will coincide with the inflammation caused there. The initial symptoms include generalized weakness and numbness of the legs, sensitivity to light, and severe headache. As the infection progresses, the symptoms advance to nausea, vomiting, and diarrhea. Children can have more severe central nervous system symptoms including coma, seizures, and paralysis, leading to a 20 percent mortality rate. In pregnant women, the fetus can be affected. Effects may include encephalitis, placental damage, spontaneous abortion, or birth defects.

No specific treatment exists except for supportive care that may include pain medications. If seizures result from the infection, anticonvulsant medications may be required. The most severe phase of the infection lasts between 24 to 72 hours and may be followed by lingering or permanent paralysis or lethargy.

Protecting yourself from blood and body fluids should be exercised to avoid developing a secondary infection. Decontamination of equipment can be accomplished using a bleach solution or by destroying the virus with heat (80°C for 30 minutes).

Quick Reference 3.8 summarizes important characteristics of VEE.

VIRAL HEMORRHAGIC FEVERS (VHF)

Viral Hemorrhagic Fevers are a group of viruses that cause uncontrollable external and internal bleeding. Since Ebola has been seen in the news many times in recent years, these types of infections have a potential to cause hysteria among members of a population. Terrorist would more likely trigger this hysteria through false threats and hoaxes rather than obtaining one of the viruses for use. The possibility of gaining possession of these viruses and then using them on a population always exists, however.

VHF are caused from viruses of several families. The viruses include Ebola and Marburg virus (Filoviridae family), Lassa fever and Argentine viruses (Arenaviridae family), Hantavirus, Congo-Crimean hemorrhagic fever virus (Bunyaviridae family), and yellow fever virus and dengue hemorrhagic fever virus (Flaviviridae family).

In general, these viruses infect and injure the blood vessels causing leakage of the blood into the surrounding tissue. The early signs of infection are fever, pain, and bleeding on the eye surface. As the infection progresses, mucous

QUICK REFERENCE 3.8 VIRAL ENCEPHALITIS (VENEZUELAN EQUINE ENCEPHALITIS – VEE)

TRANSMISSION
Mosquito borne infection in nature
Aerosol transmission during bioterrorist attack

SIGNS AND SYMPTOMS
Fever, headache, muscle pain, general weakness, then neurological disease progressing to coma

DIAGNOSIS
Blood testing

TREATMENT
None available
PREVENTION
None

VACCINE
Vaccines are available and new live attenuated viral vaccines are being actively investigated

ISOLATION
None

membranes hemorrhage, there is bleeding from the lungs, and shock results. In addition to causing bleeding, Rift Valley fever and yellow fever also affect the liver causing the skin to become jaundiced (yellow).

Congo-Crimean hemorrhagic fever and Rift Valley fever are naturally transmitted by insects. Hantavirus infection is transmitted through aerosolized rodent feces while dengue fever and yellow fever are transmitted by mosquitoes. The natural reservoirs for the Ebola and Marburg virus diseases are unknown at this time.

Depending on the involvement of the infection, and the virus responsible, the mortality rate can be as high as 90 percent. Ebola and Lassa have the highest mortality rate and the most rapid onset of symptoms.

These viruses can be transmitted by terrorists in a variety of ways but the most dangerous is by aerosolizing the agent. It is conceivable that these viruses could be used as a warfare agent but the difficulty in obtaining and cultivating such organisms makes them an unlikely weapon for non-state sponsored terrorist organizations.

Quick Reference 3.9 summarizes important characteristics of VHF.

QUICK REFERENCE 3.9 VIRAL HEMORRAGIC FEVERS—VHF (LASSA FEVER, EBOLA VIRUS, MARBURG VIRUS, YELLOW FEVER)

TRANSMISSION
Contact transmission
Aerosol in biological terrorist attack.

SIGNS AND SYMPTOMS
Weakness, fever, muscle pain, conjunctivitis, low blood pressure, hemorraging from eyes, skin, mouth or rectum
Kidney failure

DIAGNOSIS
ELISA for antigen detection

TREATMENT
Intensive supportive care may be required
Antiviral therapy with ribavirin may be useful

PREVENTION
The only licensed VHF vaccine is yellow fever vaccine.

ISOLATION
Isolation measures and blood and body fluid precautions

BIOLOGICAL TOXINS

BOTULISM (*CLOSTRIDIUM BOTULINUM* PRODUCING BOTULINUM TOXIN)

Botulinum toxin is the most toxic substance known. Although not technically feasible, a single gram of crystalline toxin dispersed for inhalation evenly over a population could kill more than one million people. To put this in perspective, botulinum toxin is about 100,000 times more toxic than the military nerve agent Sarin that was used in the terrorist attack in the Tokyo subway.

Because of the extreme toxicity of this toxin, the Soviet Union extensively researched its use as a biological weapon during the Cold War era. During more recent times, Iraq admitted to the United Nations Inspection Team in 1991 that it had conducted research in the use of botulinum toxin as a biological weapon. Four years later it was discovered that Iraq did not only research its use, but had also filled and deployed over 100 munitions containing the toxin.

Seven types of botulinum toxin are produced from the bacillus *Clostridium botulinum*, a spore-forming bacteria. The presentation of the illness related to the botulinum toxin is known as botulism.There are three presentations of botulism: foodborne, wound, and intestinal. Naturally occurring cases are rare and are usually caused from improperly prepared or canned food. Once colonized in the body, the bacteria releases the botulinum toxin systemically, causing signs and symptoms that range from visual difficulty, difficulty swallowing, and dry mouth to overall body paralysis. In order to prevent the occurrences of botulism, all food should be heated to more than 240.8°F or boiled for 10 minutes to destroy the bacterium and toxin. The occurrence of botulism is somewhat rare, but there have been outbreaks. The largest outbreak in the United States occurred in 1977 when 59 people were identified and treated after eating poorly preserved jalapeno peppers.

Botulinum toxins act by blocking nerve impulses. The interruption of nerve impulses is evidenced by descending paralysis starting at the head causing double vision, blurred vision, drooping eyelids, slurred speech, difficulty swallowing, dry mouth, and muscle weakness. The illness can progress into respiratory failure secondary to paralysis of the respiratory muscles. Therefore, definitive treatment may include fluids, nutritional support, and mechanical ventilation that may last from weeks to months.

The onset of signs and symptoms is generally quicker after ingestion of the bacterium than it is for after inhalation exposure. Depending on the dose of the toxin, the onset of symptoms from an ingestion of the bacterium can be two hours to 10 days with the average onset of 12 to 36 hours. The inhalation symptoms generally occur between 24 to 36 hours after exposure and progress to respiratory failure slower than with foodborne exposure.

An antitoxin is available through the Centers for Disease Control but may not be considered in mass casualties cases because up to nine percent of treated persons experience some hypersensitivity. The vaccine is given at intervals of 0, 2, and 12 weeks and requires an annual booster. A person taking the vaccine develops protection from the infection after the third dose. That protection then slowly diminishes, necessitating the administration of an annual booster.

Persons with botulism do not require decontamination and provide no risk to others. Botulism is not contagious and has never been reported to cause a person-to-person infection. Patients who survive from the illness may have shortness of breath and fatigue for years, as complete recovery only occurs after all affected nerves are regenerated.

Quick Reference 3.10 summarizes important characteristics of botulism.

QUICK REFERENCE 3.10 BOTULISM (*CLOSTRIDIUM BOTULINUM* TOXIN)

TRANSMISSION:
Inhalation of aerosolized toxin
Ingestion of contaminated food sources

SIGNS AND SYMPTOMS:
Inhalation and foodborne share symptoms
Blurred double vision, skeletal muscle paralysis, respiratory paralysis

DIAGNOSIS
ELISA test

TREATMENT
Equine antitoxin provided by CDC

PREVENTION
None

VACCINE
Department of Defense toxoid

ISOLATION
Not necessary

STAPHYLOCOCCAL ENTEROTOXIN B (SEB) (*STAPHYLOCOCCUS AUREUS*)

Staphylococcal Enterotoxin B (SEB) represents another form of food poisoning that incapacitates victims but rarely causes death in the process. The clinical effects are caused from toxins produced by the bacterium *Staphylococcus aureus*. The illness caused by SEB is much more common than botulism. The natural outbreaks are usually clustered and can be traced to one exposure source such as a restaurant or picnic. Terrorists may find SEB simple to produce and easy to use to contaminate open food sources and water supplies. This exposure could affect hundreds or thousands of persons, taxing local healthcare facilities and generating desired publicity. The effects from the exposure can last several weeks causing a long-term strain on the healthcare infrastructure.

The effects are different depending on the route of exposure. Inhaled SEB causes systemic injury that can lead to shock. Ingestion of SEB causes a slower onset of symptoms that are generally less dramatic and less serious. The mechanisms of toxicity are complex and involve the use of the body's own antibody response to cause injury.

The onset of symptoms occurs 3 to 12 hours after an inhalation exposure and is characterized by fever, headache, muscle pain, shortness of breath, and occasionally chest pain. A gastrointestinal exposure may also include nausea, vomiting, and diarrhea.

There is no specific treatment so only supportive care is given, including maintaining an open airway and administering oxygen.

Decontamination of exposed equipment can be accomplished using a bleach solution. Vigorous hand washing is recommended and care should be taken to not eat any food that may have been contaminated with the toxin.

Quick Reference 3.11 summarizes important characteristics of SEB.

QUICK REFERENCE 3.11 STAPHYLOCOCCAL ENTEROTOXIN B (SEB)

TRANSMISSION:
Ingestion of toxin in contaminated food
Inhalation of toxin during bioterrorist attack

SIGNS AND SYMPTOMS:
Symptoms appear in 3–12 hours after ingestion or inhalation of toxin
High fever, headache, muscle pain, chills, cough, shortness of breath, severe diarrhea, low blood pressure

DIAGNOSIS
Antigen detection in urine
Toxin assay of nasal swabs

TREATMENT
Hydration fluid and electrolyte replacement
Respiratory support

PREVENTION
None

VACCINE
None

ISOLATION
None required

RICIN

Ricin is a biological toxin formed in the seed of a castor plant (*Ricinus communis*). This toxin is important to review as castor plants are grown throughout the world and are unregulated. Castor beans are a cultivated crop and are used in the production of castor oil produced by the cold crushing of the bean. Castor oil is nontoxic and used as a cathartic. The remaining mash contains toxic levels of ricin. The ingestion of two to four beans by an adult or one to three beans by a child is enough to cause poisoning and death. Although there have been only a few single-victim events, terrorists could produce ricin as a solid and aerosolize it to affect large numbers of victims. It can also be used to contaminate food products causing injury and death to those who ingest it. Once the ricin gains access into the body it blocks protein synthesis thus killing the cell.

When inhaled, the cellular damage in the respiratory system leads to the death of tissue, bronchitis, pneumonia, and the injury allows fluid to enter the lungs. The onset is rapid with symptoms seen in as little as three hours but more commonly 8 to 24 hours. When ingested, the damage is seen in the gastrointestinal system and is evidenced through gastrointestinal bleeding. These symptoms are followed by cardiovascular and systemic injury including vascular collapse, liver and kidney damage, and death.

There is no antidote for ricin toxicity. If an exposure to this toxin is suspected or known, proper protective measures must be taken to protect eyes, open skin, and the respiratory system. If a victim is contaminated, decontamination of the patient can be accomplished by washing with soap and water.

If ricin poisoning is not suspected at autopsy, the cause of death can be easily missed. It must be specifically tested for to be found, making it an ideal weapon for terrorists wanting to kill a limited number of persons.

Quick Reference 3.12 summarizes important characteristics of ricin.

TRICHOLTHECENE MYCOTOXINS (T2)

Mycotoxins are chemically complex toxins that are naturally produced by fungi. There are more than 40 such mycotoxins that are capable of destroying the integrity of cell membranes. The toxin targets the most rapidly reproducing cells so the cells found on the skin, the mouth, rectum, hair, and in bone marrow are the most at risk.

T2 can enter through all routes, causing injury along whichever routes are exposed. Symptoms appear within minutes and are evidenced by burning, itching, reddened skin, burning in nose, throat, sneezing, burning of the eyes, and conjunctivitis (infection around the eyes). The burning and reddening rapidly advances to blackened necrosed tissue. Gastrointestinal symptoms include nausea, vomiting, diarrhea, and abdominal pain.

QUICK REFERENCE 3.12 RICIN

<u>TRANSMISSION:</u>
Ingestion, inhalation, injection

<u>SIGNS AND SYMPTOMS:</u>
Inhalation: bronchitis, pneumonia and fluid in the lungs in as little as 3 hours but more commonly 8–24 hours.
Ingestion: gastrointestinal bleeding
Followed by cardiovascular and systemic injury, vascular collapse, liver and kidney damage, and death

<u>DIAGNOSIS</u>
Special blood and tissue testing

<u>TREATMENT</u>
Supportive

<u>PREVENTION</u>
None

<u>VACCINE</u>
None

<u>ISOLATION</u>
None required

All clothing worn by a victim must be removed to lessen exposure and prevent secondary exposure. This toxin is extremely hearty requiring a heat source of 500°F for 30 minutes to be destroyed. The most efficient detoxifying agent is sodium hypochlorite (household bleach), which can eliminate the toxic activity of T2. The victim should be washed with soap and water to remove residual material—bleach or bleach solutions should not be used to wash people. T2 is not water soluble so complete decontamination will require careful, complete washing.

There is no specific antidote to T2 poisoning and supportive care has limited benefits. If T2 is ingested gastric lavage should be provided followed by ingestion of activated charcoal.

Quick Reference 3.13 summarizes important characteristics of T2.

QUICK REFERENCE 3.13 TRICHOLTHECENE MYCOTOXIN (T2)

TRANSMISSION
Ingestion, inhalation, injection

SIGNS AND SYMPTOMS
Burning, itching, reddened skin, burning in nose, throat, sneezing, burning of the eyes, and conjunctivitis (infection around the eyes)
Rapidly advances to blackened necrosed tissue
Gastrointestinal symptoms include nausea, vomiting, diarrhea, and abdominal pain

DIAGNOSIS
Blood chemistry testing, liver enzymes

TREATMENT
Supportive

PREVENTION
None

VACCINE
None

ISOLATION
None required

SUMMARY

If a terrorist chooses to use a biological agent to effect harm on a population it will probably be carried out in a covert attack. The initial clue will be a large outbreak of similar illnesses in a normally healthy population. The likelihood that a biological weapon could be used on American populations is very real, due to the ease in acquiring these substances.

It is estimated that approximately 52 different biological agents have been weaponized, although it is unknown what the exact list may be. Government militaries have genetically engineered some of the bacteria to be resistant to the more common antibiotics. There is also significant speculation that bioengineering has been used to splice several microorganisms to develop a super microorganism that is resistant to antibiotics, very infectious, and easily transmitted. This form of "black biology" is thought to have been practiced in the former Soviet Union.

CHEMICAL AGENTS

OVERVIEW

The use of chemicals to harm or kill innocent persons leaves an unsettling feeling in the minds of most civilized people. Words such as inhumane, unjust, unfair, barbarian, and cruel all come to mind. Nonetheless, militaries from all over the world have used chemical agents since even prior to World War I. Some agents remain exactly as they were first developed in those earlier days while others have gone through a metamorphosis to increase their effectiveness. It is reasonable to believe that terrorists learn how to cause destruction and death from the military—many are actually sponsored by military forces. The manners in which these warfare agents may be used in a civilian population must be understood and self-protection efforts that can be followed upon their use must be identified.

These chemicals are classified in military terms describing their effect on the enemy. It is clear that the intention of these agents is twofold: one is to incapacitate; and the other, and a much more serious effect, is to kill. In this guidebook, chemicals will be identified by their effect on the body. When the intention of the chemical exposure is to cause death, a notation will be made in the text. This information will be based on the properties of the chemicals and their toxic ratings.

For the most part, chemical agents will be broken down into nerve toxins (military "nerve agents"), chemical asphyxiants or suffocants (military "blood agents"), respiratory irritants (military "choking agents"), skin irritants and burning agents (military "blister agents") and antipersonnel chemicals (riot control agents). Because terrorists may choose to use other commonly found industrial chemicals to complete their tasks, it is important to understand that there are many other chemical possibilities that may be used as weapons in clandestine warfare. Terrorists could find it more convenient to use chemical railroad tankers, tractor-trailer trucks or other bulk storage units for their attacks, releasing large quantities of industrial chemicals into a population. Among these are anhydrous ammonia, hydrogen fluoride, hydrocyanic acid, and sulfur dioxide, to name a few. All citizens should be aware of the typical military chemical agents as well as the more common toxic gases and poisons used in industry.

NERVE TOXINS

Nerve toxins are probably the most common agents selected for use in wartime activities. These agents are very effective because they can enter through virtually any route; inhalation, ingestion, or absorption through the skin and eyes. The amount of the exposure and the toxicity of the nerve agent determine if the victim is incapacitated or dies. The military nerve agents are extremely toxic to the intended target, but are also formulated to break down rapidly in the environment so that invading troops can safely inhabit the area within days after the attack.

Many of these military nerve agents began life as commercial grade insecticides that were found to be extremely toxic to humans. Similar chemicals used today in farming industries are organophosphate pesticides—some possess alarming toxic qualities. Terrorists may choose to use commonly found industrial strength pesticides rather than military agents to injure or kill intended targets. Pesticides like parathion and tetraethylpyrophosphate (TEPP) are both commercial-grade insecticides that have toxicity levels approaching military nerve agents. These pesticides have limited availability but are easier to obtain than their military counterparts.

Whether military or civilian chemicals are the weapons of choice, the physiology of poisoning is the same. The military is able to transport and use many of these chemical weapons safely by producing two less-toxic chemicals that when mixed, become the desired toxic nerve agent. This principle can be employed by a terrorist to lessen the danger of transporting extremely toxic chemicals, which can be mixed later, at the point of delivery.

This method was used by the extremist terrorist organization the Aum Shinriko during their chemical attack in the Tokyo subway on March 20, 1995. Terrorists placed a set of bags of binary chemical in six different locations in the Tokyo subway. They then mixed the chemicals together using a stick. The combined chemicals formed the military nerve agent, Sarin. Sarin evaporates at a rate slightly faster than water. During the attack, toxic Sarin vapors filled the subway station. The results were incredible with 5,510 victims complaining of symptoms and 12 deaths. The Sarin used was estimated to be about 20 percent pure and therefore not as efficient as it was intended to be. The event was considered by the Aum Shinriko as a failure because only 12 died in the attack while the goal was to cause a massive number of deaths.

MILITARY NERVE AGENTS

Tabun (GA), Sarin (GB), Soman (GD), and VX (V-agent) are the most widely known military nerve toxins. These agents are organophosphate compounds that are very similar to industrial grade pesticides. Although most are formulated to be a respiratory poison, some have also been made thicker, with less volatility, to be used to make victims sick when their skin touches a contaminated surface.

Tabun, Sarin, and Soman (G agents or German agents) are volatile and evaporate slightly faster than water. This makes the G agents very dangerous as inhalation hazards. Terrorists can provide a heat source to these chemicals and increase the volatility, thus increasing the hazard.

None of these chemicals are easy to produce but both Tabun (GA) and Sarin (GB) are much simpler to synthesize than Soman, thus making them the terrorist's choice among military-type nerve agents. Soman, although very difficult to formulate, is the most deadly of the G agents. The chemical's poisonous action becomes permanent within about two minutes, making nerve-agent antidotes inefficient.

VX (V for Venom) is not as volatile as the G agents, only evaporating as rapidly as motor oil. This particular nerve agent's intended use is to contaminate areas and cause poisoning when unsuspecting victims rub against or touch the chemical. For VX to become a respiratory hazard it must be mechanically aerosolized or heated to increase the evaporation rate. Because of the thickness and poor evaporization rate of this agent, it tends to last much longer in the environment causing injury or death days after its release.

PHYSICAL PROPERTIES

All military nerve agents are clear, colorless, and are either odorless or have very slight odors. The less pure these agents are, the more noticeable the odor is. German agents can have a fruity odor while VX has a sulfuric odor.

For the most part, military nerve agents are found in a liquid form that evaporates at varying rates. Although the media and Hollywood continuously refer to these chemicals as nerve gas, this is a common misnomer. Although the most efficient way to introduce this poison is through the lungs, by breathing the chemical vapors as it evaporates some nerve agents are formulated to rapidly enter through the skin when they are touched. These agents are made thicker so they will not evaporate very fast, but once on the skin will be absorbed over a short period of time.

ROUTES OF ENTRY

These agents can enter through all routes but inhalation of the vapor causes the most rapid action. Because an enemy cannot guarantee that inhalation will occur, some of these nerve agents have been thickened so that they will not evaporate rapidly but will persist on objects for longer periods of time, transmitting to a victim when brushed against. Thickened Soman and VX were made specifically for this purpose. Commercial-grade organophosphate pesticides are also treated with a thickening chemical so they coat the leaves of a crop. These pesticides are formulated to last only about 24 hours and then break down into nontoxic chemicals. Both commercial-grade pesticides and military nerve agents can be used to contaminate food or water sources, causing illness through ingestion.

An exposure to vapors usually generates symptoms within seconds or minutes and can range from mild to severe. Symptoms like eye pain and blurred vision may be the first signs of poisoning. If the victim is out of the contaminated environment the symptoms will usually not progress after a couple of minutes.

Ingestion of the chemical causes symptoms within minutes or hours, then rapidly increases in severity. The illness progresses until all of the poison is either absorbed or eliminated from the body. Therefore, the illness may worsen over hours making the severity of the symptoms unpredictable.

When there is a liquid exposure on the skin, the onset of symptoms is slower than for an inhalation exposure and range from five minutes to 18 hours. This will commonly trigger muscle twitching at the site of absorption. If not removed from the skin, the symptoms can worsen over hours.

PHYSIOLOGY AND SIGNS AND SYMPTOMS

Once in the body, both military nerve agents and commercial-grade pesticides work by causing an overstimulation in a portion of the nervous system. These agents generally affect the parts of the nervous system that control functions we do not have to think about. Functions like sweating, saliva production, and urination will be overstimulated. Therefore, a victim poisoned with a nerve agent will have symptoms such as excessive salivation, urination, diarrhea, excessive mucous production, muscle twitching, and seizures. Severe pupil constriction is also common to the point of severe eye pain and blurred vision. A more complete list of symptoms is listed in Quick Reference 4.1.

QUICK REFERENCE 4.1 NERVE AGENTS (TABUN, SARIN, SOMAN, VX)

EFFECTS
Pinpoint pupils, runny nose, watery eyes, shortness of breath, and convulsions
Septicemic

ONSET
Seconds to minutes

TREATMENT
Decontamination
Atropine
2 PAM Chloride
Valium for convulsions

SIGNS AND SYMPTOMS ACRONYMS FOR NERVE AGENTS

Salivation	**S**alivation	**D**efication
Lacrimation	**L**acrimation	**U**rination
Urination	**U**rination	**M**iosis (pinpoint pupils)
Deficaiton	**D**efication	**B**ronchospasms
	Gastric pain	**E**mesis
	Emesis	**L**acrimation
		Salivation

DECONTAMINATION

The military has performed many studies on the decontamination processes for nerve agents. Water, both fresh and sea, have the ability to decontaminate nerve agents through a process of hydrolysis. This process works well for most of these military agents but because some, like VX and commercial-grade pesticides, are thickened with an oily substance to make them last longer in the environment, water alone is not adequate. The addition of an alkaline soap (like laundry or dish soap) has proven to increase the efficiency of the decontamination process. The military uses a sodium hypochlorite solution (household bleach) that works by oxidizing the nerve agent, effectively deactivating it. These bleach solutions can irritate the skin, burn the eyes, and bleach hair. For these reasons, bleach should only be used to decontaminate objects and not be used on people.

TREATMENT

Both military nerve agents and commercial-grade pesticides can be treated with antidotes. These antidotes are not usually available to the general public so definitive care must come either from Emergency Medical Services (EMS) or hospitals.

Because victims may be producing great amounts of airway mucous and saliva, breathing may be compromised. Keeping victims calm until assistance arrives is very important. Providing oxygen is one of the first treatments that should be accomplished when help arrives.

The antidotal treatment consists of two drugs; atropine and 2PAM (pralidoxime chloride, protopam chloride). The military provides these antidotes to soldiers as a Mark I antidote kit. The kit includes atropine and 2PAM in a self-administered, intramuscular injector. Because seizures are a common effect from these nerve toxins, the military also provide Valium injectors to reduce the occurrence of seizures. EMS and hospitals may choose to administer the antidotes intravenously, for faster action. Unfortunately, in mass casualties, local EMS and hospitals may not have antidotes in sufficient quantities to treat every victim.

CHEMICAL ASPHYXIANTS

Chemical asphyxiants are a group of chemical agents that have the capability of causing suffocation through chemical action in the body. This is different from simple asphyxiants that cause death by displacing oxygen from an occupied area, thereby suffocating victims. Many chemical asphyxiants are used in industry and may be easily obtained by terrorists to inflict harm on their targeted victims. The most common civilian-used chemical asphyxiant is cyanide. Cyanide is used for heat treating and plating, fumigation, and chemical synthesis in the production of plastics. It is found as a gas (hydrogen cyanide), as a solid (cyanide salt), or as a liquid and is a common component of many compounds containing carbon and nitrogen. Other chemical asphhyxiants include hydrogen sulfide, carbon monoxide, and organic nitrogen compounds (nitrates and nitrites). This class of warfare agent can be easily obtained and used to harm individuals or a large group of targets.

PHYSIOLOGY OF CHEMICAL ASPHYXIANTS

Chemical asphyxiants inflict harm by not allowing the victim to use oxygen once it enters the respiratory system. Some actually alter the blood's ability to carry oxygen while others prevent the body cells from using the oxygen. In either case, the victim is breathing in regular air with full concentrations of oxygen but the cells are not getting the oxygen they need to survive.

MILITARY BLOOD AGENTS

The military chemical asphyxiants (blood agents) consist of two chemicals, hydrogen cyanide (AC) and cyanogen chloride (CK). These agents are identical to their civilian counterparts used in industry. Their availability is one reason that terrorists may choose these agents to cause harm. Cyanide has a long history of use for both executions and homicides. Jim Jones chose cyanide to kill 900 of his followers at Jonestown, British Guiana. Seven more lost their lives in Chicago when Tylenol tablets were tainted with cyanide by a perpetrator attempting to kill his ex-wife. Even the chemist accredited with the discovery of cyanide, Karl Whelhelm Scheele, lost his life in a laboratory accident involving this chemical. Militaries throughout the world have looked to this agent as one of their primary lethal weapons.

Hydrogen cyanide is a liquid at less than 79°F, but vaporizes rapidly. As a vapor or gas it is lighter than air and rises and dissipates rapidly when used in an unenclosed environment. Cyanogen chloride is a liquid at less than 55°F and vaporizes even quicker than hydrogen cyanide. It is twice as heavy as air and thus has a tendency to linger in low-lying areas for longer periods of time.

Quick Reference 4.2 summarizes important characteristics of blood agents.

QUICK REFERENCE 4.2 BLOOD AGENTS (HYDROGEN CYANIDE, CYANOGEN CHLORIDE)

EFFECTS
Panting, convulsions, loss of consciousness, respiratory arrest

ONSET
Minutes

TREATMENT
Decontamination
Oxygen
Assist ventilations
Cyanide antidote kit

CYANIDE

Cyanide is one of the most rapid-acting poisons. It gains access into the body most often through inhalation, but can also be ingested and absorbed through the skin and eyes. It can cause death within minutes to hours, depending on the concentration, route of entry, exposure time, and activity level of the victim. The speed at which cyanide gas works is evidenced by how rapidly a death row prisoner dies during a gas chamber execution in which hydrogen cyanide is generated through the use of sodium cyanide being dropped into sulfuric acid (usually within a minute).

The victim may present with a wide variety of signs and symptoms because cyanide poisoning affects virtually all of the cells in the body. The most sensitive organ to the lack of oxygen is the brain, where the urgent need for oxygen is first sensed. Early signs can include headache, restlessness, dizziness, agitation, and confusion. Later effects are seizures, coma, and death.

Because both of the military cyanide agents vaporize rapidly, they readily mix with air and, in well-ventilated areas, disperse quickly. Since respiratory-system exposure is the most common route of entry, a cartridge mask or fire-fighter breathing device is needed if a rescue is attempted in a contaminated atmosphere.

Cyanide has an odor of bitter almonds but can only be detected by 60 to 80 percent of the general population. The ability to detect this odor is a sex-linked recessive trait, with the deficit occurring more in the male population. Three times more women can smell the odor than men.

Cyanide poisoning (hydrogen cyanide and cyanogen chloride) and hydrogen sulfide both work by inhibiting the body cells' ability to use oxygen. Following an exposure, the cyanide or sulfide is transported by the blood to the body cells. Once in the cells, these chemicals bind with an enzyme that is needed to carry oxygen through the process of making energy. The cells are immediately affected because they cannot produce the energy needed to keep the body functioning. Although the cyanide levels in the body are cut in half every hour after exposure, death may occur before the body has a chance to detoxify itself. A late sign of a cyanide poisoning is red or flushed skin. Well oxygenated blood is bright red in color. The redness of the skin is the result of the blood being completely full of oxygen and not having the ability to transfer the oxygen to the cells.

If decontamination is needed (because of exposure to liquid or solid cyanide) soap and water is recommended. Exposure to the gas will only warrant the removal of clothing.

Treatment for cyanide poisoning is tricky and time dependent. Since respiratory arrest develops quickly, if the victim gets a significant exposure, providing artificial breathing is immediately needed. Cyanide poisoning can be treated with an antidote kit containing three pharmaceuticals. The kit contains amyl nitrite, sodium nitrite, and sodium thiosulfate. For the antidote to be effective, it must be started early in the poisoning. These kits are carried by some EMS providers and at hospitals though in small quantities. If mass casualties related to cyanide exposures occur, the kits may not be available in sufficient quantities to help all victims.

The patient who survives an initial exposure must be closely monitored and admitted into the hospital. This precaution is necessary because of the late complications that may develop affecting breathing and blood circulation

CARBON MONOXIDE

Carbon monoxide actually bonds to hemoglobin, which is that part of the blood that normally carries oxygen. This bond between carbon monoxide and hemoglobin occupies the place where oxygen normally bonds to the hemoglobin. Therefore, oxygen cannot be carried on those areas occupied by the carbon monoxide. This causes suffocation even though the victim may be in an atmosphere with full levels of oxygen. This condition leads to asphyxia and death. When carbon monoxide is bonded to the hemoglobin it makes the blood appear very red in color. This red color is even brighter than when oxygen bonds with the hemoglobin. Therefore, one of the late signs of carbon monoxide poisoning is "cherry red" skin. There is no antidote for this poisoning so the treatment for carbon monoxide poisoning consists of supplying 100 percent oxygen and supporting breathing.

NITRATES AND NITRITES

Nitrate/nitrite poisoning also involves the hemoglobin but in a different way. These poisons actually change the chemical makeup of hemoglobin so it will not pick up and transport oxygen. The new compound is called methemoglobin. Methemoglobin can be changed back to hemoglobin with the antidote, methylene blue. Methylene blue is administered only in hospitals and is commonly found in hospital pharmacies. Unlike cyanide and carbon monoxide, the reddening of the skin does not occur in nitrate poisoning. Instead, the lips and skin appear blue. This is because methemoglobin is brown in color and changes the color appearance of the skin where blood is near the surface.

In either of the above cases of poisoning (carbon monoxide or nitrate/nitrite), death is imminent once 60 to 70 percent of the hemoglobin is out of service for the transportation of oxygen. Unconscious victims must get immediate treatment using 100 percent oxygen for successful resuscitation and, when an antidote is available, it must be made ready as soon as the victim reaches the hospital.

RESPIRATORY IRRITANTS

Strong respiratory irritants have a long history of use by military forces during wartime activities. Documents about World War I are filled with incidents of chemical warfare involving chlorine and phosgene gases. These gases remain today in military arsenals around the world. When discussing terrorism, these agents cannot be discounted because of their prevalence as industrial chemicals.

CHLORINE, PHOSGENE, AND ANHYDROUS AMMONIA

Many communities use chlorine gas in the pure form for chlorinating drinking water. It is also used as an antimold and fungicide agent. Compounds containing chlorine are even used for chlorinating home swimming pools and cleaning toilets and showers. Chlorine-based products have a very distinctive odor common around public swimming pools.

Phosgene, although not as common as chlorine, is also found in industry. It is used for organic synthesis during the production of polyurethane, insecticides, and dyes. It is also a byproduct of burning freon, which has led to many injuries among firefighters. Phosgene has a sweet odor and is said to smell like new-mown hay.

Although not thought of as a military agent, anhydrous ammonia also fits into this category because of its ability to cause severe respiratory irritation and injury. Anhydrous ammonia is a common industrial refrigerant and is used in the blueprinting process. Anhydrous ammonia smells just like the ammonia bought for home cleaning, an odor that is unmistakable in higher concentrations.

Even at low concentrations, these chemicals released into a crowd could send multitudes of patients to the hospital and trigger a panic scenario. These reactions may fulfill the needs of a terrorist.

MILITARY CHOKING AGENTS

Phosgene (CG) and chlorine (Cl), have been in military arsenals for almost a century—though they are no longer used on the battlefield. Both of these agents are typically stored as liquids but rapidly become gases once released into the atmosphere. Their expansion ratios allow them to be transported in small containers and, once released, fill a much larger area.

By using these agents on the battlefield, the goal was to incapacitate the enemy so that they could be overrun by the advancing troops. This strategy worked well because both of these gases once released on the battlefield were rapidly dispersed by winds and environmental conditions but did not leave contaminated objects behind to cause injury to the troops arriving later.

Once exposed, victims are overcome with severe, uncontrollable coughing, gagging, and tightness in the chest to the point of stopping respiration. Other injuries include burning of the eyes, skin, and nose. All of these effects are very unpleasant and could cause panic and death at large gatherings or in heavily populated areas.

Phosgene, in particular, is poorly soluble in water and tends to cause deeper lung injury. These deep lung injuries allow the fluid from the blood to leak into the airways, filling the lungs with fluid. This can occur hours or days after an exposure. The victim of a severe exposure literally drowns in his own body fluids. When the injury is not fatal, the victim is left with destroyed lung tissue and a lifetime of respiratory illnesses.

Injuries to the upper part of the respiratory areas are usually a result of an exposure to those chemicals that are more water soluble, like chlorine and ammonia. These chemicals readily dissolve into the moisture-coated airways. In the case of chlorine, this results in the production of hydrochloric acid and chemical burns in the air passageways. After an ammonia exposure, an alkali is formed that also causes a severe long-lasting burn injury. Both chlorine and ammonia cause swelling of the airways and asthmalike symptoms.

The military choking agents are heavier than air, giving them the ability to find low-lying areas and persist for longer periods of time. This was the case during World War I when the troops would dive into foxholes and trenches for cover from gunfire only to choke from one of these respiratory irritants. In the civilian world, hazardous-materials responders have been dealing with these chemicals for years. The typical response includes using well-placed hose lines toward the cloud of gas to disperse it using a combination of air currents and water absorption.

Decontamination with water is only necessary when victims are complaining of burning skin. Flushing eyes with water may also help to reduce the burning experienced by a victim. The removal of clothing may be all that is necessary to effectively reduce the hazard. All of these chemicals are gases and will disperse rapidly into the environment.

The treatment for respiratory irritant exposure is to provide oxygen. Asthma medicines meant to dilate airway passages or reduce swelling are also effective in allowing more air to pass through swollen airways. These treatments should be provided by EMS or other healthcare providers so victims may be rapidly treated and transported to the hospital.

Quick Reference 4.3 summarizes important characteristics of choking agents.

QUICK REFERENCE 4.3 CHOKING AGENTS (PHOSGENE, CHLORINE, AND AMMONIA)

EFFECTS
Tightness in chest, coughing, and shortness of breath

ONSET
Minutes to hours

TREATMENT
Decontamination
Bronchodilators
Assist ventilations

SKIN IRRITANTS AND BURNING AGENTS

HYDROCHLORIC ACID AND ALKALIS

Many chemicals are capable of burning the skin and eyes. Typically this injury is related to an acid or alkali being splashed on a person. These chemicals are used to a great extent in industry and even at home. Hydrochloric acid is used in swimming pools and spas to balance the pH of the water. Alkalis are used as degreasers, oven cleaners, and drain openers. All of these chemicals are capable of injuring people if not used or handled with great care.

Though industrial chemicals are capable of causing skin irritation, none do so to the degree that military blister agents can. It is for this reason that any discussion in this book about skin irritants will center on the military agents.

MILITARY BLISTER AGENTS, (VESICANTS)

The military primarily uses three types of blister agents. These agents include mustard (H) and related variations of mustard (HD, HN, and HT), phosgene oxime (CX), and lewisite (L). All of these agents are liquids with the exception of phosgene oxime. These agents vaporize slowly to cause an inhalation hazard. Skin and eye burning and irritation are the most common effects that result from direct contact with the liquid.

Military blister agents were originally developed because airborne respiratory irritants were affected so much by wind and other environmental conditions that their use was limited. Enemy troops also adapted quickly to airborne hazards by developing masks that would protect them from chemicals like chlorine and phosgene. Blister agents were therefore primarily directed at skin and eye exposure but could also affect the respiratory system when the intended victim is not protected. These agents are very harmful in the liquid form but also vaporize and become an airborne contaminant as well. Most of these chemicals date back to World War I, some having been further refined through the years to become even more efficient.

Mustard was developed during World War I and continues to be a major chemical warfare agent since that time. This agent was reportedly used in the 1960s by Egypt against Yemen and was used again during the 1980s as a weapon between Iran and Iraq. The United States still has stockpiles of this agent in both Colorado and Utah. The containers that store the mustard stockpiles are breaking down and leaks from these containers are common. The United States is currently working to incinerate these weapons.

These types of agents are strong irritants capable of causing extensive burns, extreme pain, and large blisters on contact. If vapors are inhaled, lung tissue will respond like the skin and form large obstructing blisters. Once the blisters break, large open wounds result, providing the opportunity for overwhelming infections and eventually death.

After mustard gains access into the body it is absorbed into the water of the tissue. Once exposed, a victim may not have any symptoms for some time and not be aware that an exposure took place. Two to 24 hours later, the mustard reaction becomes evident with the formation of blisters on the skin, death of tissue in the airways, and severe irritation to the eyes including swelling and temporary blindness. The eye injury can lead to permanent scarring and loss of sight.

Lewisite acts very similar to mustard but has additional effects throughout the body. Symptoms beyond the blistering effect may include breathing difficulty, diarrhea, vomiting, weakness, and low blood pressure. The eye burns resulting from an exposure to liquid Lewisite are devastating and if not decontaminated within one minute the damage will probably be irreversible.

Mustard has a freezing point of 57°F so it solidifies at temperatures less than this. For this reason pure mustard may not be a good choice in colder climates. Mustard also vaporizes slowly making it primarily a skin-contact hazard. It takes temperatures of greater than 100°F for enough of the liquid to vaporize and become a respiratory hazard.

Both lewisite and phosgene oxime vaporize more readily than Mustard, making them more of a respiratory hazard. All of the blister agent vapors are heavier than air allowing them to stay near the ground and not dissipate quickly. Lewisite vapors are seven times heavier than air so they will persist in low-lying areas for long periods of time.

Decontamination for all of the blister agents must be immediate. Each one of these agents effect harm to tissues upon contact. Mustard is different than the other agents, however, as it does not trigger symptoms for several hours, leaving the victim without a clue that an exposure took place. Lewisite and phosgene oxime both cause irritation and pain almost immediately—this alerts the victim about the exposure, stimulating a quicker decontamination process. If a victim notices liquid on the skin, decontamination should be accomplished by first blotting off the agent, taking care not to blot with contaminated material. If the agent is wiped instead of blotted, the agent is spread the length of the wiped area, extending the injury.

The decontamination solution of choice is soap and water. The United States military suggests that this decontamination process be followed with a bleach solution but this is only suggested for contaminated objects and not people.

Quick Reference 4.4 summarizes important characteristics of blister agents.

QUICK REFERENCE 4.4 BLISTER AGENTS (MUSTARD, LEWISITE, PHOSGENE OXIME)

EFFECTS:
Red and swollen skin, blisters, eye irritation, blindness, shortness of breath, coughing

ONSET:
Minutes to hours

TREATMENT
Decontamination
Topical antibiotics
Bronchodilators
British Anti-Lewisite

RIOT-CONTROL AGENTS (LACRIMATING AGENTS)

These agents are used to incapacitate the victims and are not intended to kill. The civilian use of these agents includes riot control and self-protection. Chemical antipersonnel weapons have gained popularity in both the general public and law enforcement because they are able to subdue persons without the use of extraordinary physical force. These chemical sprays offer a nonlethal form of protection that causes temporary extreme discomfort. Generally, there are three versions of these sprays available: chloracetephenone (CN), orthochlorobenalmalononitrile (CS), and the most popular civilian agent, oleoresin capsicum (OC).

CN

The effects from this chemical agent begin in one to three seconds and are characterized by extreme irritation to the eyes causing burning and tearing. Irritation to the skin is also common because the CS crystals stick to moist skin, causing burning and itching at the point of contact. CN also irritates the upper respiratory airways, causing coughing and burning of the throat. These effects generally subside after 0 to 30 minutes and usually require no medical assistance.

CS

Symptoms start in about three to seven seconds and last 10 to 30 minutes. The effects are stinging of the skin especially in the moist areas and intense eye irritation with profuse tearing. The burning also affects the nose and throat. Some victims panic due to the feeling of shortness of breath and chest tightness. Victims describe effects as being ten times worse than those of CN. Some police agencies still use this irritant mostly for crowd dispersal.

Both CN and CS are solid submicron (less than 1 micron) particles. They are extremely light and are carried to the target area in a carrier solution that evaporates quickly, dispersing the agent. Because of the light, fine particle, both of these chemicals are prone to cross-contamination between the initial victim and anyone attempting to assist them.

OC

OC has become the safest and most popular of the chemical agents. It is found in police aerosol sprays and over-the-counter agents. It is a non-water soluble agent prepared from an extract of the cayenne pepper plant. The contact of OC causes immediate nerve-ending stimulation, burning pain, and excessive tearing. The effects from OC start almost immediately when contact occurs and lasts about 10 to 30 minutes. The exposure usually leaves no lasting effects.

TREATMENT

Less than one percent of the exposed victims will have symptoms severe enough to need medical care. As mentioned earlier, the symptoms generally last only about 30 minutes and leave without lasting injury.

There is no known antidote for these irritants so medical care is centered on relief of the symptoms. Flushing the eyes with water will help but the pain will return as soon as the irrigation stops. The pain only ceases after the chemical has been decontaminated through natural body processes.

DECONTAMINATION

Decontamination should consist of removing the clothing and washing exposed areas with soap and water. In the case of OC exposure, the agent is not water soluble so the effects on mucous membranes will last until they are detoxified by the body and not washed away through irrigation. This detoxification takes place over about 30 minutes, which is the duration of symptoms. Irrigation of the eyes is usually not necessary but will relieve the burning pain.

SUMMARY

The use of chemicals against a population is a cruel way of attacking anyone but is particularly inhumane when used against innocent persons. Unfortunately, most military chemicals have a civilian counterpart that is commonly used in industry. Chemicals like chlorine, ammonia, and phosgene are used in industry and their use against a population could be easily carried out with devastating results. The use of nerve and blister agents causes injuries that are more severe and more challenging to control as the extreme toxicity of these chemicals is not matched in industry.

Reasonable replacements for nerve agents are industrial pesticides such as parathion. Both the military nerve agents such as Sarin, and industrial pesticides like parathion, fall into the organophosphate chemical family. Both cause similar physiological effects. The major difference is that the exposure required to get the same results is much less for the military grade nerve agent.

Strong alkalis like sodium hydroxide (lye) can cause injury much like military blister agents such as mustard, but again, the tremendous amount of tissue devastation that a military agent can cause cannot be achieved using an industrial strength chemical.

Industrial chemicals are much easier to acquire and are very prevalent in our society. Most, if not all of them, are loaded into trucks and travel down our highways making the chemicals extremely vulnerable to sabotage during point-to point-delivery. Many of the same chemicals are carried in quantities of 50,000 to 60,000 gallons in railroad cars that travel through major cities along rail lines or are parked in transfer yards located in close proximity to concentrated populations.

Industrial chemicals present a great hazard to populations if terrorists choose to use them. Efforts to increase the security of these chemicals have been taken while other plans are still underway. Even so, there remains a threat that military-grade chemical warfare agents could be either manufactured in this country or imported from a state-sponsored terrorist organization for use against a population here in the United States A working knowledge and an understanding of how these chemicals affect a population is the best defense that someone can have to protect themselves if an attack takes place.

5

BOMBINGS

OVERVIEW

Although the threat of using chemicals or biological agents is on the rise, the "tried and true" expression of terrorist activity is the bomb. During acts of terrorism, bombings are used about 70 percent of the time. There are several reasons that explosive devices are used so often. First of all, they are cheap. The bomb used by Timothy McVeigh at the Oklahoma federal building cost under $500.00. Not only is the cost attractive but so is the availability. McVeigh's bomb was made with the fertilizer ammonium nitrate. This product can be bought in 50 pound bags from most garden supply centers. The fertilizer was then mixed with fuel oil and detonated with a secondary detonation device and a lit fuse.

Another reason that bombings are chosen is the instant recognition that a bombing brings. The instant that a bomb is detonated, public fear and media attention begin. For terrorists it is as important (or even more important) to get this kind of attention as it is to cause harm and devastation. This instantaneous effect does not occur with the use of a biological agent although the harm and fear caused can be many times greater.

Well before the 1998 Olympics in Atlanta, the expectation of a violent attack was being discussed and preparations were made. Every worker and security person hired for the summer Olympics was instructed to look for anything out of the ordinary that might be a suspicious device. At a pre-Olympic concert in Centennial Park a security guard noticed a satchel lying on the ground in the crowd. He realized that the satchel was an object out of place and began clearing the crowd away from the area when the bomb exploded. The explosion killed one person and injured many others. Ironically, though the security guard was truly a hero, he was wrongly accused of planting the bomb.

PHYSIOLOGY OF A BOMBING

Every person should be aware of their surroundings, especially at large gatherings. If you see or suspect something is out of the ordinary or unusual, trust your feelings and move away. The further from a blast a person is, the better their chances of survival. A large blast can generate a positive wave pressure of 6,000 pounds per square inch. This pressure wave would immediately disintegrate a human body. In fact, even a wave of 100 pounds per square inch can cause death. The wave rapidly decreases in pressure the further away a person is from the point of detonation. Pressures as low as a half pound per square inch can rupture eardrums and cause a loss of balance. Table 5.1 describes both the potential injuries and structural effects from an explosion.

Table 5.1 Blast Injuries and Structural Effects

Potential Injury	Pressure (psi)	Structural Effects
Off balance, Rupture of eardrums	0.5–3 psi	Glass shatters, facade failure
Internal organ damage	5–6 psi	Cinderblock shatters, steel structures fail, containers collapse, utility poles fall
Pressure causes multisystem trauma	15 psi	Structure failure of typical construction
Lung collapse	30 psi	Reinforced construction failure
Fatal injuries	100 psi	Structural failure

When an explosion occurs, the blast creates a pressure wave that extends in all directions away from the blast. This wave of pressure is followed by a vacuum created by this wave. Persons in close proximity to the blast are initially thrown away from the explosion by the positive wave. After that wave passes, a body can be almost immediately drawn back toward the blast by the vacuum. The result is an injury similar to what is experienced in an automobile accident and is termed an acceleration/deceleration injury. A body being thrown one way then the other tends to tear internal organs and cause severe traumatic injuries.

The size of the bomb also gives some clues as to the characteristics of an explosion and how to protect oneself. In fact, the Department of Alcohol, Tobacco, and Firearms (ATF) formulated both an evacuation distance and lethal blast distance. These distances are calculated for car and truck bombs and are based on size. Table 5.2 is representative of the chart developed by the ATF.

Table 5.2 Lethal Blast and Evacuation Distance

Vehicle Type	Maximum Explosive Capacity	Lethal Air Blast Range	Maximum Evacuation Distance	Falling Glass Hazard
Compact	500 lbs.	80 ft.	1,500 ft.	1,250 ft.
Full Size	1,000 lbs.	125 ft.	1,750 ft.	1,750 ft.
Passenger or Cargo	4,000 lbs.	200 ft.	2,700 ft.	2,750 ft.
Small Box Van (14 ft.)	10,000 lbs.	300 ft.	3,750 ft.	3,750 ft.
Box van or Water/Fuel Truck	30,000 lbs.	450 ft.	6,500 ft.	6,500 ft.
Semi Trailer	60,000 lbs.	600 ft.	7,000 ft.	7,000 ft.

BOMB THREATS

Bomb threats are one of the most difficult situations to handle safely. The vast majority of time, bomb threats are perpetrated to cause disruption and not to actually warn anyone of a pending explosion. In the 1970s and 1980s, bomb threat hoaxes became an everyday occurrence in cities across America. From schools to grocery stores, all were victims of these hoaxes—some were even done as practical jokes. Although these threats can never be ignored they can be downplayed to a point that the perpetrator does not get the attention they want. In most, if not all, cases of major bombings there is no warning beforehand. Instead, the bomb is covertly planted in or near the target.

If a bomb threat is received, the person receiving the call must remain calm. Demonstrating panic on the phone will probably stimulate the person threatening the bombing to call in other threats. The person receiving the call should get whatever information they can from the caller. Information about the caller's accent, any background noises, where and what type of device has been placed, and the time frame for detonation are all important clues that should be assessed. The person receiving the call should be instructed to ask where the device is located. Do not be surprised if the caller answers all of these questions. Real bombers are extremely proud of their devices and some will even brag about how it is designed and what it will take to detonate it.

Unfortunately, in most cases real bombings will not be accompanied with a warning. For the most part, arbitrary phone calls warning of a bomb without any specific information are probably hoaxes. Those that give very specific information, however, should be considered credible and warrant immediate action.

If a phone call advising of a bomb is received, immediately notify the authorities. Do not make this call using a cell phone. If a cell phone must be used, leave the immediate area to make the call. Cell phones and radios can trigger the detonation of a radio-controlled bomb.

If your bomb-threat plans call for you to evacuate, walk around before leaving your work area, and quickly look for anything out of the ordinary. Usually the police and fire departments will not search as they are not knowledgeable enough about your workplace to know if anything is suspicious. They will rely on you for that information and will want you to advise them if anything is out of place.

When you and your office workers evacuate, make sure all of you gather at a predetermined muster point well away from the building. For high-rise structures a safe distance is defined by a distance away from the building equal to the height of the building. Quick Reference 5.1 summarizes important information as to what to do in case of a bomb threat.

QUICK REFERENCE 5.1 BOMB THREATS

- In the case of a bomb threat, get whatever information you can from the caller; accents, background noises, where and what type of device, time frame for detonation.
- In most cases a real bomb will not be accompanied with a warning.
- Notify the authorities—DO NOT USE CELL PHONES.
- Before leaving your work area quickly walk around and look for anything out of the ordinary. Usually the police and fire departments will not search—they will rely on you to do that function and will want you to advise them if anything is suspicious.

SECONDARY DEVICES

In 1996 and 1997, there were numerous bombings at abortion clinics and gay bars around Atlanta. In two of these cases, the initial bomb exploded and a secondary device was planted to explode once emergency responders arrived. At one of the abortion clinics a secondary device exploded next to a fire department command car. The device caused only minor injuries because of the placement but the intent of the device was clear, to kill or injure emergency responders. The other secondary device, found outside of a gay bar, did not detonate—it was discovered and defused by the local bomb squad.

Although the use of a secondary device in this country is relatively new, it is a tactic that has been used in Northern Ireland for many years. In fact, bombers in Northern Ireland have taken this principle one step further. In a more recent bombing, a threat was called into a business causing an evacuation of the residents to a nearby street corner. After enough time had gone by without an explosion it was assumed to be a hoax and the occupants returned to work. A couple of days later a new threat was called into the building and again the occupants evacuated to a nearby street corner. That time a car bomb was parked at the street corner and exploded killing many more people than would have been killed if a bomb was placed inside the building.

Another illustrative example of terrorists drawing people into danger took place at an American embassy. The embassy bombing in Nairobi, Kenya was so successful in killing the occupants because a small primary device was used first and then was followed by a large secondary device. The perpetrator first threw several hand grenades next to the building. This drew the occupants of the embassy to the windows to see what was occurring outside. When a larger truck bomb exploded, the loss of life and the number of injuries were huge. The death toll reached 257 and about 5,000 people were injured.

These secondary devices teach a valuable lesson. That lesson teaches both emergency responders and bystanders that being in the proximity of an earlier explosion can be very dangerous. When an explosion occurs, leave the area immediately. Do not stay around and look or attempt to help. Allow the scene to be handled by professional responders wearing protective equipment who are aware of the unique hazards presented after a bombing.

INTERNET BOMB-MAKING INSTRUCTION

The Internet is a wonderful medium for researching and discovering new and interesting subjects. Unfortunately, it also provides the means to find ways to build and develop bombs. This was the case in Littleton, Colorado when a couple of high school kids decided to attack schoolmates and teachers. These kids were able to build very large satchel bombs that contained propane cylinders, flammable liquids, and pipe bombs with a timing device and an electric fuse. They were also able to assemble homemade hand grenades made with CO2 pellet gun cartridges, gunpowder, and self-striking matches. The self-striking matches would ignite the fuse of these grenades as they bounced down the hallways of the high school. Many printed Internet pages involving bomb-making information were found in the perpetrators' rooms after the incident.

SURVIVING AN ATTACK

OVERVIEW

We have seen over and over again as terrorist bombings resulted in the collapse of occupied buildings. Although many occupants die as a result, many others become trapped in the rubble hoping that they are found before time runs out. As a prepared person, there are some things that you can do to increase your survivability if a building collapse takes place and you are involved.

When the Oklahoma City bombing caused part of the Murrah building to collapse, within minutes the rubble was being swarmed with onlookers wanting to help the victims. Those who reacted without training in an effort to help caused additional confusion and mayhem in an already disastrous situation. Many times the rescuers refer to these self-appointed helpers as expedient responders. These expedient responders, even with the best intentions, use up the immediate professional resources as they arrive on the scene of a disaster. One of the first challenges to the professional emergency responders during the Oklahoma event was gaining control of the scene. This took precious time away from real rescue.

HOW TO SURVIVE A BOMBING AND BUILDING COLLAPSE

IF YOU ARE A VICTIM

So what should you do if you are a victim or in close proximity to a structure bombing? We can learn a lot from those who have survived such events. First, if you are able to get away from the building on your own, do so and do so quickly. Once a building has been damaged by a blast it becomes very unstable and may continue to collapse for days. Therefore, immediate evacuation will increase your survivability. Second, as you leave the disaster area, help any other walking wounded that you can but do not stop for unconscious or trapped individuals—they will be helped by those emergency rescue workers trained and equipped to function in dangerous circumstances. The key is to get away quickly. Once away from the scene, stay away and do not go back.

Quick Reference 6.1 summarizes what to do in case of an explosion.

QUICK REFERENCE 6.1 EXPLOSIONS

- Leave the area immediately.
- If items are falling, head for a doorway, exit as soon as possible.
- If you are trapped in the debris, cover your mouth and nose.
- Use a whistle or tap on metal or concrete, do not shout—this can result in inhalation of dust (wait until the air is clear).
- When evacuating, take other walking victims with you. Lightly trapped individuals can be assisted also.
- Leave unconsciousness and trapped—they will be helped by professional rescuers.

IF YOU ARE TRAPPED

If you become trapped, there are some things to do to help you get rescued. First, make noise—lots of noise. Bang metal, concrete, scream (if the air is clear), whistle, or anything else you can do to alert the rescuers that you are there. Second, try to stay calm and do not panic. If days are needed in the rescue effort, you must sustain for that period of time. Panic burns up needed energy and significantly reduces survivability. If you have a cell phone with you, use it. Call 911, call a friend, a family member, or anyone who can help the emergency responders locate you in the rubble.

YOUR SURVIVAL KIT

After witnessing the events that can take place as the result of buildings being bombed, it is prudent to be ready for the unexpected. To be ready, it is helpful to carry a survival kit any time that you attend large group meetings, sporting events, conduct business in government buildings, or go to work, especially if you work in a multilevel building. These kits can be easily assembled and carried on your person. This survival kit should include:

- A strong, good quality, folding knife. A knife is the universal tool. It can be used to signal, cut clothing or fabric to make bandages, and has a variety of other uses.
- A whistle (lifeguard style). These can be purchased in most boating stores. The high pitch shrill produced by a good whistle can be heard for great distances.

- A small flashlight with fresh batteries. A penlight will also work. Laser pointers are now inexpensive but should be carried in addition to, and not as a replacement for, a regular flashlight. These items can alert rescuers. If direct line of sight can be established with the rescuers, a laser pointer can produce a beam of light for a very long distance even in the sunlight.
- A good-fitting particle mask. Particle masks are made of either cloth fabric or paper. Fitted properly these masks will keep you from breathing large amounts of dust but do not protect against any toxic or poisonous gases.
- Twenty feet of cord or light, strong rope. Cord can be used for a variety of reasons including to signal rescuers or as a tool.
- A 24"x24" piece of brightly colored plastic (preferably red or orange). This plastic can be used as a signal flag or be pushed from an opening to alert rescuers.
- A cell phone. Cell phones are becoming a standard for most Americans. In a disaster, cell phones are invaluable. Directing a rescue worker to your exact location cannot be done in an easier way than with a cell phone.

All of these items can be placed in a ziplock sandwich bag and be carried in a purse or pocket. If you know you will be going to a high-risk event such as a political convention or international sporting event, in addition to the survival kit, carry a bottle of water. Today, many people carry their own bottled water for convenience, and are unknowingly better prepared for a disastrous situation. If you were to be trapped in a building collapse, a bottle of water could sustain you for days beyond what could be done without it. A person can generally live without food for seven to nine days but without water for only two to four days.

HOW TO SURVIVE A CHEMICAL OR BIOLOGICAL ATTACK

LETTERS AND PACKAGES

Following the events in September 2001, Americans experienced a new threat—anthrax-laced letters. Although for years hoax letters were sent all around the country earning the name "the bomb threat of the 90s" it was not until October of 2001 that the real threat was experienced for the first time. As a result, many actions were taken to prevent people from opening suspicious letters and packages. Below are the clues that should be used to determine if a letter or package is suspicious. Flyers containing this information were sent from the FBI and Postal Service to companies all over the country. A representation of this flyer is located in Figure 6.1.

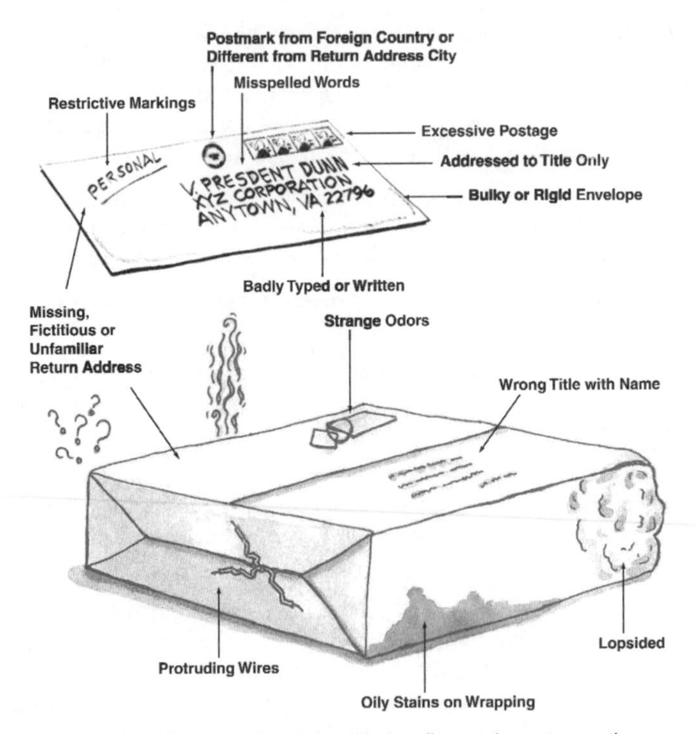

Figure 6–1 *Ensure safety while handling and opening mail.*

What else can be done to ensure safety while handling and opening mail? The list below will help guide you in the safe handling of mail:

- Latex gloves. Use latex gloves when handling bulk mail. Gloves are generally not needed when handling your own mail unless there has been an instance of an anthrax letter found in your area and there is a possibility of your letters being sorted at the same location as the anthrax letter was sorted. Gloves will prevent anthrax spores from entering any open wounds that may be present. In at least one case of cutaneous (skin) anthrax, an open wound on the chest was infected by the transference of spores from the hands to the wound.
- N95 mask. This mask is used in both industry and healthcare. It is capable of filtering out very small particles including anthrax spores, preventing inhalation anthrax. If you work in a business that receives large amounts of mail and the threat of receiving anthrax letters is a possibility, then the use of an N95 mask should be an option. Fit testing for the mask is required to insure that air is not allowed to enter around the mask. It is uncomfortable to wear for long periods and can cause some respiratory symptoms in persons having existing respiratory illnesses such as emphysema.

- Letter opener. Using a letter opener to open all letters would reduce the spread of any substance contained inside. One way people open letters is by tearing off a corner then inserting a finger and ripping open the top in an outward pull, toward the body. Others open letters by tearing off the end then blowing into the envelope to spread the sides (Johnny Carson style) and expose the letter inside. By far the safest way to open a letter is with the use of a letter opener. Once inserted under the flap a letter opener can be pushed away from the body and face. If anthrax or any other contaminate is inside the letter, it is pushed away from the face and not easily aerosolized.

Quick Reference 6.2 summarizes what to do in case you receive a suspicious package.

QUICK REFERENCE 6.2 SUSPICIOUS PACKAGES

- Do not touch a suspicious package.
- If you have handled it or are holding it put it down carefully and immediately wash your hands then call authorities.
- If you have opened a letter which at first glance was not suspicious but has become questionable put it down, cover it with a piece of paper and immediately wash your hands.
- Leave the room and shut the door but stay in the area and remain calm.
- Immediately wash your hands and blow your nose.
- Call for the authorities.

If something suspicious is suspected in a letter or package, the safest action is to not move it. If you are holding the item, carefully lay it down, leave the room, close the door, and immediately wash your hands. Then call authorities. In many of the anthrax-letter cases, the letter was carried from one office to the next causing contamination in multiple places in the building.

CHEMICAL VAPORS

Letters are not the only means to spread biological or chemical agents and cause terror. In fact, a letter is the least efficient way to spread these dangerous agents. If is far easier to spread a hazardous substance in the air over a targeted population, contaminate objects, or poison water and food. The most efficient means of causing harm to a large population is to place the substance in the air while the least efficient is to contaminate objects.

Aerosolized chemicals can be used to cause harm in a number of ways. First, a liquid that evaporates at a steady rate can be placed in an enclosed area and left to vaporize into an airborne poison. This was the tactic used in the Tokyo subway when Sarin was placed in six locations in the subway. The results were devastating with 12 deaths and 5,500 injuries.

The second is to leak a gas and use the wind to move the plume over the population. This can be very effective when the gas is significantly heavier than air and the environmental conditions are just right. An example of this action was found when Sadam Hussein used bombs filled with a nerve agent against the Kurds. These bombs exploded near the ground upwind of the village using the wind to carry the poisonous vapor into the village. Hundreds of Kurds were killed as a result of the attack.

If you are warned or suspect the use of a chemical vaporized agent outside or near your location, move to an inside room. If you have identified a safe room in your home for other disasters, it can be used for this purpose also. Close all doors and windows and turn off the air-conditioning or heating system. Have a radio or television available in safe room so you can tune into the news and know when it is safe to leave. Information on the safe room is located in Chapter 8.

CHEMICAL LIQUIDS AND SOLIDS

Chemical liquids and solids are also effective when used to cause injury to a population. Highly toxic chemicals can be used to contaminate an environment frequented by people. This concept has been used by militaries with blister agents that were thickened with the addition of a heavy motor oil-type substance. This ensures that the chemical will not vaporize quickly and allows it to stick to objects and not run off. Chemical liquids can also be covertly sprayed over a group. One possibility for this type of dispersion is using a mechanism that is not viewed as suspicious. In an effort to keep large populations cool during summertime outdoor activities, many times large fans are used to keep the crowd cool. These disperse a fine mist of water into the wind created by the fans, thus providing a cooling sensation to the crowd as they walk by. Inventive terrorists can use processes like this to disseminate chemicals onto an unsuspecting crowd. By using a chemical that has a delayed response, thousands may become victims before the first even complains.

Chemicals can also be used in solid form. Many of today's pesticides are manufactured as a solid particle. These chemicals can be disseminated using cropduster aircraft or by blowing the solids into the air using an explosive device or pressurized sprayer. These fine powders can be carried by the wind for long distances and contaminate those in the path of the plume created by the chemical. Dry agents like these are activated by moisture and therefore can remain full strength until they stick on moist skin, becoming hydrated, and begin causing harm.

If you are aware of an attack of this type in your area, the most important action you can take to remain safe is to limit your travel around the community. These agents can float in the air for long periods and move around with wind currents. By staying in your home and away from the affected area you will limit the chances of becoming another victim. If you are exposed, immediately remove your clothing. Get into a shower and wash with soap and water. If this is not possible, most Emergency Medical Services and fire departments have been trained to decontaminate victims.

AIRBORNE BIOLOGICAL AGENTS

For the most part, biological warfare agents were intended to be blown into the air. The military found that this means of dispersal is very efficient, if the intent is to injure or kill a large population. Biological agents in the form of a fine powder or mist released upwind from a population can be devastating. In fact, in 1970, the World Health Organization assessed both the dissemination of plague and anthrax upwind of a city of five million and found that between 36,000 to 100,000 would die while many more would become sick, overwhelming both emergency services and healthcare systems.

Biological agents are unlike chemicals in that the onset of symptoms usually occur days to weeks later. To the contrary, symptoms from a chemical exposure range from immediate to usually no longer than a couple of hours. The long onset of illness presents some challenges to a community that has experienced an outbreak related to the covert spread of a disease-causing organism.

Until the location of the contamination is discovered, the community should be warned to stay away from areas where people gather. Malls, airports, churches, and other places where large groups of people assemble should be suspect and avoided until determined to be clean. Also, the thought of a second attempt should not be ruled out.

Remember that the perfect target for terrorists is a large assembly of people. By no means can you, or should you, avoid every gathering that occurs, however. To do so would not be prudent. We must be able to assess the mood and temperament of terrorist activities in our country and prepare based on a real or perceived threat.

FOOD- AND WATERBORNE BIOLOGICAL AGENTS

Efforts are currently being made to improve the security of our public water supply. Even without that additional security, contamination of the public water supply by biological agents would be difficult. In most communities water is treated with some type of chlorinating system. Chlorine is one of the best and most efficient ways to kill biological microorganisms. Today's treated water contains concentrations of chlorine that are high enough to kill most of these agents. Contamination of water is more likely to occur closer to the end user, such as in water supplies going into buildings, occupancies, or restaurants. These would probably not be arbitrary attacks but directed at the business type or the population served by the business.

Biological agents used in food pose a higher risk and their effects approach those seen after an airborne distribution of a biological agent. Contaminating food that is not normally cooked is an effective way of causing harm to large populations. The best way to avoid becoming a victim of contaminated food is to avoid eating from open food bars. Salad bars and food bars are usually not closely watched and it is very easy for someone meaning harm to contaminate it with biological agents.

TRAVEL SAFETY

OVERVIEW

Previous acts of violence perpetrated against Americans increases our focus on travel safety. There is no ideal way to ensure your safety while traveling, especially while traveling outside of this country, but there are some simple tips to follow to give you an edge. During high-hazard times (when the threat of violence is high), every effort should be made to postpone travel plans until the threat is reduced. The news media will advise when it is considered safer to travel. Organizations like AAA can also be accessed to provide travel information concerning the safety of traveling outside of the country.

TRANSPORTATION

Traveling in your car is one of those activities in which you should become more aware of your surroundings—more than just looking at traffic patterns and using defensive driving techniques. Being aware of events other then typical traffic can be especially important during times of heightened awareness. As mentioned earlier, the use of typical chemical haulers on the highway and railways is an option for terrorists. Always avoid vapor clouds and liquids coming from bulk chemical trucks. If a cloud or vapor is in your path, do not attempt to travel through it. In these cases, the product being released from the container can displace the surrounding air, lowering the oxygen concentration to the point that your car's engine will stop running, stalling your vehicle in the vapor cloud. This was the case a number of years ago in Memphis, Tennessee when a propane truck was involved in an accident, releasing the contents into the air. Many vehicles attempted to drive through the cloud, which caused their engines to stall. The eventual fireball created when the product was ignited killed everyone who had attempted to drive through it but had become stranded in the cloud. Although this was not an intentional attack, the results were devastating.

If you are in your car and the radio advises you of a chemical or biological dissemination, stay in your car, close the windows, drive away from the identified threat, and turn your air conditioning/heating off or place it in the recycle mode. In your car you should have disaster supplies which would allow you several hours of safe time while you are sheltered there. These supplies should include food in the form of snacks, water, a particle mask, and cell phone. Citizen band radios are also a good asset. Many of these items are already carried by those people who live in the northern climates as they are subject to getting stuck in severe snow storms and these emergency supplies would be needed if they become trapped.

TRAVELING ABROAD

It is unfortunate that recent terrorist events play a part in governing our desire to travel and experience other destinations and cultures. In some cases, even though our desire might not change, travel to certain destinations may be restricted. These restrictions on American travel are created by other countries as well as our own. Cuba has long been a destination that has restrictions on travel from the U.S. Generally, traveling to much of the Middle East is risky and dangerous. Although there is much history and interesting culture to explore at these locations, the decision to travel to these areas must be considered seriously. Carefully plan those locations you want to visit and always focus on your own safety as a priority.

The State Department issues travel warnings which are based upon reliable political, social, and economic variables. They recommend which countries to avoid and can issue avoidance bulletins or Consular Information Sheets. Consular Information Sheets include items such as the location of the U.S. Embassy and information about the consulate, immigration information, regulations, currency, health conditions and controls, legal penalties, and the political environment. Furthermore, U.S. embassies in travel destinations can be contacted for further warning information.

In some cases, a warning is not issued but rather a safety or security condition is stated. This is a summary of stated facts about the situation in the desired country of travel and in no way gives permission or advice on the travel—it just gives information on the conditions of the country.

At times, public safety announcements are disseminated to give short-term information about the country and the perceived conditions of travel within that country. For example, if political unrest is viewed as a possible threat, this information is released to warn Americans of a possible coup, threats, violence, social unrest, etc. These are transnational conditions that can cause a threat to the traveler.

As a general rule before making your travel plans abroad, contact the State Department for warning advisories and public announcements on the country in question. In light of the current world situation, the State Department may either restrict or strongly advise against entry into a variety of countries.

FLYING

Today, the thought of flying makes many people fearful. Because of the recent events of terrorism, passengers are more on edge and pay close attention to their surroundings. Baggage is checked in a much more methodical fashion and each passenger is screened with greater diligence. There are a few things, however, that you can do to make the traveling more efficient and avoid delays.

PLANNING AHEAD

- Ask your family and coworkers to not reveal your plans to strangers.
- Get all paperwork and documentation ready in advance. Passports, visas, and other travel documentation should be in hand and ready prior to the flight.
- Make sure that your travel itinerary is known by both your family and, if traveling on business, your office. If the itinerary changes make sure you update them.
- Make sure that your destination knows your itinerary. If on business, your destination contact should know your flight number and time of arrival. Additionally they should have a contact name and number to call if you do not arrive as scheduled.
- If traveling abroad make sure that all vaccinations and immunizations are up-to-date.

ITEMS TO LEAVE AT HOME

- If you work for what could be perceived as a controversial company, do not allow any company logo items to be seen. Logos from government offices, defense industries, religious organizations, and others that could make you stand out in a crowd should be packed away in your checked luggage or left at home.

- Membership cards that could be misconstrued and used against you in a hostage situation are dangerous. Items identifying you as belonging to a religious or governmental organization, or even your company identification cards should be left behind.
- Expensive jewelry should be left behind. This includes gold watches, large rings, and other items that attract unreasonable attention.
- Take only the credit cards you need for the trip. Large wallets with numerous credit cards attract attention and may attract criminal activity.
- Anything that resembles a weapon should not be carried. Even a small pocketknife can be construed as a weapon and may cause unexpected delays at airport check-in. If you normally carry a pocket knife, pack it in your checked luggage prior to arriving at the airport. Remember September 11th hijackings were perpetrated using box cutters.
- A complete copy of your travel plans, including daily plans while out of town should be left at home or sent/faxed to your point of contact at your destination.

ITEMS TO TAKE WITH YOU

- Pack an ample supply of prescription medications, clearly labeled, preferably with the original pharmacy label.
- Include first aid supplies such as self-adhesive bandages, antiseptic and nonprescription pain medicines.
- Carry with you a complete list of all medications and allergies. This can be on a card in your wallet so it can be easily discovered in the case of an accident or sudden illness.
- Use plainly colored and nondescript luggage.
- Pack your business cards in your checked-in luggage.
- In some countries you must have a camera or photo permit. If you plan to take pictures in a foreign country check with a travel agent or the country's travel bureau for requirements prior to packing your camera.
- Carry a drivers license. You can receive an international drivers license from the American Automobile Association (AAA).
- Take travelers checks that can be exchanged for local currency once at your destination.
- Carry telephone numbers and addresses of your destination contact.

- International credit cards such as American Express, VISA, MasterCard, or Diners Club.
- Pack a photocopy of your passport and/or visa stored in an alternate location such as briefcase or checked luggage.
- Take 10 to 20 passport photographs of yourself. You may need these to get passes or foreign identification cards.

SUMMARY

For many Americans, the freedom of travel has been taken for granted. The recent events of terrorism have taken the feeling of safety away and replaced it with concern and travel anxiety. These feelings have made travelers more aware of their surroundings and left them with a desire to take an active role in their own safety. By taking some simple, common-sense steps, your travel can be greatly improved and much safer then ever in the past.

DISASTER PLANNING

OVERVIEW

Any occurrence in one's life that is unforeseen is generally an incident that causes a high degree of turmoil. These events produce anxiety, which can lead to fear. Fear of the unknown and the anxiety of wondering what should or should not be done produces more worry and stress. Fear is a choice with which we live. If we choose to deal with it, we must take steps to prepare for the event we fear, whatever that event may be.

For example, people living in the northern United States have a high probability of experiencing severe snowstorms and they prepare for such an event. In the southern states, floods and hurricanes are an annual event and again the population prepares for when the threat becomes substantiated. For those in the Midwest, tornadoes are the threat and communities consistently become prepared to deal with them. If people all over the country prepare for such disasters, why would we not prepare for an event that is intentional, covertly preplanned with the intention to kill or injure large populations, and usually comes without warning?

Preparation is a function that reduces fear by taking proactive steps to protect yourself. As with any emergency, if you understand some basic principles and take an active role in your own preparedness, the sense of control will give you higher levels of confidence and comfort when an incident occurs. Preparation can be the difference between your family's health, safety, and preservation, and the anxiety, despair, and uncertainty that an event may produce.

RISK ASSESSMENT

By evaluating a situation and understanding the possible ramifications of some subtle observations, you can make some logical and important decisions. For example, if you were to notice dead animals, an unusual number of insects, or an area-wide wilting of plants you can make the assumption that a chemical or biological agent is the cause. These signs may occur as the result of an accident, a natural occurring event, or an intentional release with the intent to cause harm. Ignoring these signs and failing to react can be devastating. In today's political environment, and considering past experiences of acts of violence against Americans, these clues should lead you to be suspicious. Be very suspicious. Refer to Quick Reference 8.1 for more information on warning signs and appropriate reactions.

The recent events of contaminated letters being sent through the mail has altered the thinking pattern of those who handle bulk mail. This new cautious way of thinking should be part of our everyday life and not just be triggered by suspicious letters. We have, and for good reason, taken our own safety for granted. Now all of us must make sure that we are more acutely aware of our surroundings so that we can take some control over our own safety.

As with any dangerous scene, your initial observations should be comprehensive, providing you with enough information to provide a risk assessment so that proper actions can be taken. It is up to you to act upon these recognizable clues. If the clues are identified in an appropriate manner then actions that alert emergency responders can be done early enough to save lives and attend to injuries.

QUICK REFERENCE 8.1 PHYSICAL INDICATORS

Look for physical indicators or warning signs such as:
- Clouds or smoke without obvious reason
- Dead animals, insects, and/or vegetation
- Disrupted infrastructure without cause (utilities, transportation, phones, etc.)
- Unusual odors, colored smoke, or vapor clouds

Take appropriate immediate action:
- Stay away from the immediate area
- Move away in an upwind and uphill direction
- Evacuate toward the predesignated safe areas (rooms)
- Notify the appropriate authorities

THE FAMILY DISASTER PLAN

Information about what kind of events to prepare for is the key to planning ahead. As previously stated, the region in which you live gives you clues about what type of emergency for which to plan. For example, those living in the southeastern states should plan for hurricanes, in the midwest states for tornadoes, and in California for earthquakes. Regardless of where you live, plans for fires breaking out in your house should be made, especially if children are possible victims. In the unfortunate event of a fire, it is important to have exit response plans for the family. In addition to these types of events, every family should have an all-hazards family disaster plan that can be used for any emergency.

The all-hazards family disaster plan should encompass all the emergencies that are possible including acts of terrorism and hazards that may arise from a terrorist-type event. We have witnessed the public hysteria that has occurred related to events of biological and chemical terrorism. After the Tokyo event and the more recent anthrax mailings, individuals went on buying sprees to purchase gas masks and chemical protective clothing. This, at first glance, seems to be a simple solution to the problem. However, in the case of a gas mask, one must realize that when a biological agent is released, you must have the mask on ahead of time for it to protect you. Furthermore, it must be the correct mask for the event or you may have no protection against the particular biological agent released. For the mask to work effectively after a chemical dispersal, it must be rated for the chemical being used. Also, each mask and skin protection ensemble must be test-fitted to the wearer. Just buying a "gas mask" may give the purchaser a false sense of protection which may ultimately cause injury. Having a gas mask is not practical even though it may be comforting.

ESTABLISHING SAFE ROOMS

For biological, chemical, or even radiological isotopes to cause harm, these agents must reach you and invade your body in some fashion. This leads us to a very practical approach toward self and family preservation. This can be accomplished by a process that is termed by emergency response professionals as "shelter in place."

As a defense against possible threats, sheltering in place can save your life even if your house or apartment is close to the affected area. All the chemicals and biological agents that we have described in this book require a certain degree of concentration before harm occurs. By staying within your house or apartment, sealing the bottom door jam with damp towels, turning off the air conditioning or heating system fans and exhaust fans, and shutting all windows, you have just created a barrier between you and the threat. Typically the threat will dissipate over time so by providing this immediate protection and staying in this protected area until it is safe to exit, you will have kept your family safe.

Once inside your safe room, stay inside and listen to the radio or television broadcasts providing information about the event. The most reliable information comes from emergency services and emergency management. By taking this action you have effectively been empowered to prepare your family against the threat. Additional planning should include having sheets of plastic that you have precut to apply to all the windows and the door of one room of the house or for the entire house. This room (or your house) becomes your safe room.

The purpose of this safe room is to provide a space where you and your family can survive a high-concentration assault of chemical, biological and/or radioactive particles. This room can be one of several places in your house, the basement, an interior room, or a first floor room with limited exposure to the exterior walls–avoid upper-floor rooms as fire or chemicals can be placed to prevent escape from higher floors. Precutting plastic sheeting and taping these sheets against window frames and door jams helps maintain an insulated barrier against the threat.

This concept has been taught for years by the federal government to residents located around the military biological and chemical storage facilities. Residents in these areas attend classes, provided through the federal government, on sheltering-in-place and are provided with kits containing plastic sheeting and tape to quickly make a safe room within their houses. Their warning comes from a loud air siren located in the middle of town. Through disaster planning and drills, all of the residents are prepared for an accidental or intentional release of these warfare agents.

DESIGNING A DISASTER LIST

In order to have an effective plan one must first design their own disaster list (see Quick Reference 8.2). This list should include food, water, and heating and cooking supplies, personal hygiene supplies, and communication tools (cell phones, citizen band radios, handheld radios) for a minimum of three days and a maximum of one week should also be included. Although being prepared represents an initial cost for supplies, it is prudent to have an all-hazards plan that includes materials to protect your home from acts of terrorism.

We can last for a week or longer without food but for only a short time without water. Your body's need for water is dependent on many things. First, the temperature plays a major role. If it is hot outside and your air conditioner is not running, the temperature inside your home will rise—causing you to sweat more in an attempt to stay cool. In these cases you will need more water to keep your body hydrated. Second, the more physical exertion you experience, the more your body needs water. Physical exertion heats your body and increases your breathing rate. Both of these contribute to loss of water. Water is the substance of life and even more so if you are planning for children, nursing moth-

QUICK REFERENCE 8.2 GENERAL DISASTER LIST

- Water: 2 gallons per day per person, 3- to 7- day supply
- Food: 3- to 7- day supply
- Clothes
- Pillows and blankets
- Toiletries
- Flashlights, radio, and batteries
- Cell phone
- Special items for infants, elderly, special needs
- Medicines
- Toys, books, and games for children
- Cash
- Vehicles with a full tank of fuel
- Duct tape and plastic sheeting
- Important legal documents
- Cash

ers, elderly and people with chronic illness. A good plan is to have at least one gallon of water for drinking per day per person. You should plan to double this requirement for personal hygiene and cooking. Do not plan to conserve water from consumption but water can be conserved when cooking, washing and for personal hygiene needs.

Dependening on the source of your water, you may have to treat the water before consumption. If you have preplanned and have water stores within your home, you may not be faced with the issue of having to treat it. However, if you rely on well water, lake water, or the threat includes contaminating public water systems, treatment of your water supply is suggested. For every one gallon eight drops of unscented bleach should be added. If the water is cloudy due to storage time or during the gathering phase, sixteen drops of unscented bleach is required. This process will kill any possible contamination from a biological agent and even work for some chemical contamination. If the water has an odor of chemicals do not consume it. Rely on the water in either canned food or bottled water.

FOOD AND WATER

Food should have nutrient value, and a long shelf life. Minimal cooking may be an additional requirement. For example, within your safe room (or sealed home), excessive cooking with petroleum gels such as Sterno can create a hazardous environment from carbon monoxide gas generated from the burning fuel. In this type of situation, cooking is not practical and can actually be detrimental. Foodstuffs should be limited to those requiring easy preparation without the need for extensive cooking.

If children are a part of your life, planning to have snacks serves to reduce their anxiety and makes the event resemble a game of indoor camping. Such food also assists in the mental well-being of the adults. Vitamins, medicines, and first aid supplies, as well as easily stored foodstuffs, should be purchased and stored before the event. Being prepared does not mean going out to buy supplies just as the event unfolds, as this adds to the concerns and anxiety felt by your family members. Planning ahead of time by buying food that you would normally eat and is simple to keep is important. Additionally, the military has perfected meals that can be stored for long periods of time, are well balanced, and come in large varieties. These meals, called "MRE" or "meals ready-to-eat", are available at most army-navy supply stores. However, these meals sometimes have an unpleasant taste and consistency. Living on them for a couple of days may be an adventure on its own.

Do not forget paper plates, napkins, and cups. It is better to throw these away then to clean them using your precious water supply. Remember that it may be days before the local emergency services can assist you. Aluminum foil for a pan barrier if cooking is also an option. Having a supply of heavy-duty self sealing bags to keep nonperishable foods fresh is also important. Your kit should also include trash bags and nonelectric can openers.

HEATING AND COOKING

Cooking and heating is somewhat difficult, especially if electrical service has been interrupted. As stated previously, if a safe room has been established, the use of fuels for heating and cooking within your sealed area may be prohibitive. All nonelectric cooking and heating sources require petroleum-based products and will create certain levels of carbon monoxide and carbon dioxide. Although cooking fuels like Sterno create low levels of carbon monoxide and are somewhat safe in enclosed areas, other fuels like charcoal or any other fuel that allows smoldering to occur, must be avoided. Carbon monoxide is generated through incomplete combustion and charcoal is the perfect example of incomplete combustion. The use of smoldering wood or charcoal will be deadly in an unventilated room. Both carbon monoxide and carbon dioxide become dangerous in an enclosed environment so fuels that generate these gases should be avoided.

If a cold environment is experienced, keeping warm should be accomplished through the use of blankets or layered clothing. If power is available then space heaters and electric blankets can be used. Keeping lights on will also provide a means of heating.

PERSONAL HYGIENE

Water stores can be used to provide personal hygiene through sponge baths. Showers and regular baths consume large quantities of water and if there is a possibility of the public water system being effected or there is a loss of power, the water system may not be available. If this is one area that you feel is important for psychological comfort, you must plan for the large consumption of water. For example the typical short shower consumes 25 to 35 gallons of water. A toilet flushes 1.5 to three gallons per use. Hand washing takes one to two gallons per wash. Brushing teeth also uses water equaling approximately one cup. Remember that we are planning for extreme conditions and that the stores of water are necessary to remain healthy. Use the stored water you have consciously and with a great deal of concern. If there is one area of planning that will assist you in your decision making process, it is in the area of water consumption and use. Become extremely frivolous with the water for all uses other then hydration.

Your disaster plan list should include toilet paper, paper towels for washing, feminine hygiene products, and trash bags for disposal. Additionally, you will have to identify a place in your safe room area for bathroom or toilet facilities. If water supplies are not interrupted, then normal hygiene can be accomplished using the bathroom facilities. However, be prepared for the worst and create a plan that also considers interruption in basic services such as water, sewage, and electricity.

COMMUNICATIONS AND INFORMATION

Knowing what is occurring in the outside world becomes extremely important in times of emergencies. Communication links to emergency response personnel and the community can be as simple as a radio news broadcasts or hand-held walkie talkies to communicate with a neighbor. Battery-powered TVs are also available and can operate if electrical service is affected. If you currently use wireless phones, have a back-up, hard line-type phone in case the power is interrupted. Regular hard-wired phones often work even when the electrical service is interrupted. Emergency radios, TVs, and walkie talkies all require batteries so preplanning for your energy consumption will be vital to your informational needs. A well-charged cell phone for emergencies during the disaster may be a consideration. Again, this is dependent on the level of infrastructure remaining after the incident.

All communities have certain radio stations which are designated as the emergency broadcast authority (if you live in a rural area a station that will simultaneously broadcast these programs are available) and are designated as the news groups that inform the public in times of emergencies. These news groups should be monitored for information about the event and instructions you and your family can take to stay safe. Local emergency management will use this means of communication to keep you informed. Instructions on when and how to evacuate, directions to shelters, how to access medical needs, and other pertinent support information can be accessed using local radio or television stations. Each broadcast is specific to the emergency within the specific community.

EMERGENCY CONTACTS

Develop a list of emergency contacts. The list should include friends and relatives that are located far enough from your local area so they are not affected by the event. These contacts can be called in times of an emergency and notified that you are alright or that you need help. These contacts should be in a different location and act as "trackers" for your family. If family members are separated due to the event, this contact can become the focal point for information dissemination to other family members and friends wanting to account for and track other family members.

In your emergency plan, establish a meeting place for family members to go to if disaster strikes. The tracker should have all of the contact numbers for that meeting place so the first family members to get there can talk to the tracker to see if contact from other family members has been made. The meeting place should be out of the immediate geographical area of your home or business. You must insist that all members of the family meet at this predesignated location if an incident takes place while the family is separated (ie. work, school, shopping, etc). This can be at a friend's or relative's house in a neighboring city or in the rural outskirts of a city. All family members should know where the meeting place is and be familiar with different routes that may be taken to get there. If communication is possible with family members after a disaster, they should be paged with a code or called and advised to go to the meeting place rather then attempt to go home. Unfortunately, in large events, phones and pagers may not work because of an overload on the phone system. In these cases, if there is even a perceived threat in returning home, the family should know to go to the meeting place.

EVACUATIONS

In certain conditions, the authorities may choose to evacuate a portion of the community. During these situations it will become important that a family plan has been developed. In mass evacuation, water, food, blankets, and clothes are not always provided. Your stock of resources may have to be a part of your evacuation plan. Additionally, not all individuals within your community may be as prepared as you are so your preparedness efforts will provide more resources to those victims that were not prepared or were unable to obtain their own resources.

COMMUNITY DISASTER PLANNING

Fire departments and law enforcement agencies across the country have been planning for large-scale disasters for many years. Part of their preparation efforts has been the establishment of trained civilian neighborhood groups called Community Emergency Response Teams (CERTs). These teams were predominately designed to deal with natural disasters that effect entire communities and overwhelm emergency services. These teams are trained to deal with the effects of tornadoes, earthquakes, and hurricanes. In communities where these teams have been trained, CERT members have become very active and effective during the preplanning and action phases of a community emergency.

CERT is a neighborhood initiative aimed at instructing a group of active citizens to provide a level of self-help during an emergency affecting the neighborhood or the community. It is designed to assist those in need when the public emergency services are crucially stressed due to the magnitude of the emergency. The goal of developing a CERT is to allow the neighborhood members to coordinate the immediate emergency functions that are required after disaster strikes and to be self-sufficient, using the talents within the neighborhood to control and deal with the issues that arise from the emergency. Each group establishes a network of individuals that can assist one another in their time of need. Within this group, a team leader (incident commander) is assigned and is responsible for funneling information and actions to the team.

Using this concept, a neighborhood can work together to become prepared for acts of terrorism affecting their community. Together they can purchase supplies and pool resources to serve not just one family but an entire neighborhood. With an all-hazards approach, these resources can be available not just in response to a terrorist event but in case of any natural or manmade disaster. Additionally, when evacuation or shelter-in-place procedures are put into place, the CERT members can take care of those neighbors that are disabled, elderly, or cannot for some reason take care of themselves. These actions and responsibilities taken on by CERT members greatly assist the local emergency services with vital functions when their own resources are stretched too thin.

As with neighborhood watch programs, everyone must become more aware of his or her surroundings. Identifying situations that seem out of place and reporting them to the appropriate authorities may actually prevent an intentional disaster from occurring. As citizens become more knowledgeable about the tactics that terrorists use, there will be less of a chance for terrorists to carry out these violent acts successfully. By changing your way of life and awareness of your surroundings you may effectively save your life and others. If something looks out of place be suspicious and investigate cautiously. When a situation occurs, take a deep breath and act appropriately. In most cases the appropriate action is to quickly leave the immediate area. If emergency responders ask or order an evacuation, do not gather your belongings, ask a lot of additional questions, or take your time preparing to leave. Just leave. Material items can be replaced but the health and well-being of you and your family cannot be.

DISASTER PLANNING IN THE WORKPLACE

We spend a great deal of time in the workplace. It is important to develop and practice an emergency plan, as you should at home. Plans addressing fires, bomb threats, suspicious packages or letters, and evacuation processes should be written down officially and emphasized. Not unlike your home emergency plans, your place of business should have plans that address issues that affect your work area, floor, or section of the building. Establishing a network to funnel information toward one team leader is important. The exchange of information between businesses, floors, or sections of the building can assist in shelter-in-place or evacuation decisions. This process has been perfected in many hospitals with the use of an incident-command system and various emergency plans to guide employees as to how to protect patients during an emergency.

The emergency workplace plans should identify safe rooms in areas that are secured and are within the interior of the building. Because you may not know when the event may occur, failure to have pre-established plans to protect the workplace will take precious time and may contribute to the injury of workers. In communities that have military chemical depots in close proximity to schools and other public-assembly buildings, emergency plans are already in place and practiced regularly. The ability to alter return air dampers in ventilation systems has been put into place so that with a few quick actions the building can be sealed from a hazardous-material leak occurring outside. These same types of action can be established in almost any building by planning ahead.

FIRE PLANNING

Fire education procedures should become a part of all employees' behavior. Make it a practice to know where the fire exits are in every building you enter and in every hotel or motel you stay in. This is second nature to firefighters that have been on the job for a number of years. Most can tell you, at any given time or place, where the exits are in any building they enter. It is also important to have a primary and secondary route preplanned. On an aircraft, know your primary exit, which should be the closest, and a secondary one if the first is blocked or you are otherwise unable to reach it.

Fire evacuation plans are designed to allow occupants self-removal from the building in times of emergency. The plan must include what is needed to immediately evacuate an area into a safe place outside of the structure. The plan may also include the movement of persons to a safe area within the structure if complete evacuation is not possible. Each section or floor of the building should have predesignated muster points outside of the building to which evacuated occupants must report. This will assist authorities and the supervisors of the business to account for the occupants. This is vitally important to fire department responders who will want to know immediately if a rescue must be made. Your local fire department can assist your planning endeavors. Quick Reference 8.3 gives more information on fire planning.

QUICK REFERENCE 8.3 FIRE PLAN

- If you are in the immediate fire area, stay low to the floor and exit the area as quickly as possible—DO NOT GATHER YOUR BELONGINGS; LEAVE
- Stay below the smoke and head towards the exit
- Cover your nose and mouth with a wet cloth
- As you leave close all of the doors around you—this will hinder the spread of the fire and allow more time for people to evacuate
- If a door is closed that you need to pass through, feel the door for heat—if it is hot, DO NOT OPEN IT but instead go toward your secondary means of exit
- Exit immediately and go to muster area

Bomb threats and suspicious packages should also be a part of the plan. For information on these items and what to do if an explosion occurs refer to Chapter 6, Surviving an Attack.

WHAT NOT TO DO IN PREPARATION FOR A TERRORIST ATTACK

DO NOT PURCHASE MILITARY GAS MASKS

Although on the surface purchasing military gas masks seems like an appropriate action, one must realize the obvious—you must have the mask on prior to the event. It does not make sense, however, to wear a gas mask every time you want to go outside for a walk, take out the garbage, or travel to work. Additionally, these masks must be coupled with a protective ensemble to protect the skin. For true respiratory protection these masks must be the right ones for the specific hazard and be fit-tested to a person's face to provide any level of protection. A mask designed for tear gas may not work for Sarin or Cyanide. None of them will work for simple asphyxiant atmospheres like carbon dioxide or nitrogen. Therefore they are not practical for the population in general. What is practical for the general population is an understanding of the possible dissemination possibilities of both biological and chemical agents. Protective measures should include critical observation to recognize possible terrorist events, use of safe rooms, and evacuation.

DO NOT BUILD A BOMB SHELTER

During the 1950s, many individuals around the country built underground bomb shelters to provide some safety from fallout after a nuclear attack. This never occurred, so these shelters became elaborate game rooms, storage areas, or were filled in. It was always questionable as to the effectiveness of this practice and how safe they really would keep those who retreated to one. Some believe that the companies who sold or built the shelters used the hysteria of Americans to make money and were not concerned about the effectiveness of the shelters.

As far as building shelters for chemical or biological attacks, the same logic holds true. They are impractical, have questionable effectiveness, and could be very expensive. It is more realistic and practical to take an all-hazards approach for emergency preparedness with your family and community. Having an emergency plan, supplies, and a safe room within your existing home is optimal.

DO NOT PURCHASE A FIREARM

The purpose of terrorism is to attack without warning. It is not conventional warfare. Instead, an attack is usually a time-limited conflict which occurs within minutes. The attack does not present a traditional bad guy—good guy scenario. Instead, it is carried out with the intention of placing a community or the country in economic, political, or social turmoil. There is generally no one to shoot at and no one to raise arms toward. In the September 11 incidents, those that were directly responsible died along with the innocent victims. The Oklahoma City bombing was a situation in which the perpetrator fled before the destruction occurred. Weapons for the purpose of protecting one's person or family will generally serve no use during a real terrorist event.

DO NOT STOCKPILE ANTIBIOTICS

This is a dangerous practice and one that could lead to serious social implications. Taking arbitrary medications for possible exposures can cause a number of negative effects. First, each bacterial infection calls for a different formulation of antibiotics. One infection is different from another requiring different doses and types of antibiotic. Additionally, viruses are unaffected by antibiotics.

The medicines in the stockpile may be the wrong type, dose, or may have even expired by the time an attack occurs. If consumed at the time of the event, there is a very good possibility that the medicine will be of no help and the false thought that some level of protection has been gained could be dangerous.

Another problem is that arbitrarily taking antibiotics can lead to the strengthening of a bacterial strain. This is what has occurred with tuberculosis. For many years tuberculosis has been a disease occurring in the American population. Because of the hardiness of these bacteria, the antibiotics must be taken for a long duration. Unfortunately, many of the population with the disease take the antibiotic until they feel better but not long enough to kill the bacteria. As the infection is passed from person to person the bacteria becomes a stronger strain because it is treated with too small of a dose of antibiotic to kill it. Instead, the bacteria becomes resistant to the treatment. With time, it takes more and more medicine to kill the bacteria and some forms cannot even be killed using traditional medications. If individuals start to stockpile their own antibiotics and take these medicines for every sniffle, cough, and flu, the effectiveness of this mode of treatment would decrease significantly.

DO NOT PURCHASE CHEMICAL-RESISTANT CLOTHES

Like the gas mask, chemical-resistant clothes are not very practical. These suits require a high level of training, support, and physical ability to use properly. Additionally, the suits must be donned prior to an exposure to a toxic chemical in order to provide any protection.

SUMMARY

Emergency preparedness and disaster planning should be accomplished using common sense and not with the mentality that we are planning for protection during an extended war effort. Terrorism is not war but instead consists of arbitrary acts of violence against Americans in an effort to bring attention to a cause, political agenda, or religious belief. Your best protection from acts of terrorism is knowledge and awareness. This book is written to give you both. Preparedness efforts should be done with an all-hazards approach and plans should address all disastrous events that could affect you and your family. Refer to Quick Reference 8.4 for a list of items suggested for an all-hazard disaster plan. Simple protection is the best action. The more complex and cumbersome an effort is, the less likely it will be successful. Be aware, act smart, and stay safe.

QUICK REFERENCE 8.4
THE COMPLETE LIST OF HOME ALL-HAZARD PREPAREDNESS ITEMS

- Assorted canned foods
- Water: 2 gallons per day per person
- Spices and flavorings
- Sweeteners
- Multivitamins
- Special needs food, baby formula
- Tools and supplies
- Battery operated radio with seven-day supply of batteries
- Flashlight with seven-day supply of batteries
- Nonelectric can opener
- Plastic storage containers
- Duct tape
- Matches in waterproof container
- Fire extinguisher
- Cash
- Aluminum foil
- Needles and thread
- Plastic sheeting
- Unscented bleach
- First aid kit
- Insect repellent
- Iodine
- Hydrogen peroxide
- Medicine dropper
- Dish-washing detergent (antibacterial)
- Hand-washing soap (antibacterial)
- Sponges
- Trash bags
- Ziplock bags assorted
- Paper towels, plates and cups
- Plastic utensils
- Water filters and or water filtration system (purification)
- Feminine hygiene products
- Toilet paper
- Dental care products
- Air freshener
- Personal care hygiene products
- Clothing and bedding
- One change of clothes per person
- Sturdy shoes
- Hat, gloves, thermal underwear
- Rain gear
- Blankets
- Sleeping bags

REFERENCES

1. Kortepeter M, Christopher G, Cieslak T, et al. *U.S. Army Medical Research Institute of Infectious Diseases' Medical Management of Biological Casualties Handbook*. Frederick, MD: Fort Detrick. 2001.

2. Inglesby TV, Henderson DA, Bartlet JG, Ascher MS, et al. Anthrax as a Biological Weapon, Medical and Public Health Management, *JAMA*. 1999; 281:1735-1745.

3. American Academy of Pediatrics, *2000 Red Book: Report of the Committee on Infectious Diseases*. 25th ed. Elk Grove Village, IL. American Academy of Pediatrics; 2000; 168-170, 212-214, 450-452.

4. World Health Organization. Health Aspects of Chemical and Biological Weapons, Report of a WHO Group of Consultants. Geneva, Switzerland: World Health Organization; 1970.

5. Chin J. *Control of Communicable Diseases Manual*. 17th ed. Washington D.C.: American Public Health Association; 2000; 20-25, 70-75, 381-387, 455-457.

6. Inglesby TV, Dennis DT, Henderson DA, et al. Plague as a Biological Weapon, Medical and Public Health Management, *JAMA*. 2000; 283:2281-2290.

7. Arnon SS, Schechter R, Inglesby TV, et al. Botulinum Toxin as a Biological Weapon, Medical and Public Health Management, *JAMA*. 2001; 285:1059-1070.

8. Shapiro RL, Hatheway C, Swerdlow DL. Botulism in the United States: A clinical and epidemiologic review. *Ann Intern med*. 1998; 129:221-8.

9. Shapiro RL, Hatheway C, Becher J, Swerdlow DL. Botulism surveillance and emergency response: a public health strategy for a global challenge. *JAMA*. 1997; 278: 433-435.

10. Henderson DA, Inglesby TV, Bartlet JG, et al. Smallpox as a Biological Weapon, Medical and Public Health Management, *JAMA*. 1999; 281:2127-2137.

11. Suriano R. Bioterror tough to carry out, experts say. Orlando Sentinal. September 30, 2001; A1, A6.

12. Angulo FJ, Botulism. In: Evans AS, Brachman PS, eds. *Bacterial Infections of Humans*, New York: Plenum, 1998.

13. Bevelacqua AS, Stilp RH, *Terrorism Handbook for Operations Responders*. Albany, NY: Delmar Publishers; 1998.

14. English F, Cundiff Y, Malone D, Pfeiffer A. Bioterrorism Readiness Plan: A Template for Healthcare Facilities, APIC Bioterrorism Task Force: 1999.

15. Stilp RS, Bevelacqua AS. *Emergency Medical Response to Hazardous Materials Incidents*. Albany, NY: Delmar Publishers;1997.

APPENDIX A

Biological Agent Quick Reference Guide: Bacteria

Bacteria	Transmission	Precautions	Treatment	Isolation	Protective Measures
Anthrax *Bacillus anthracis*	Inhalation of spores; cutaneous through skin contact; gastro-intestinal through contaminated food	Prevent contact with spores in the air-borne configuration; noncontagious between humans	Decontaminate hands and area Penicillin and Streptomycin or Doxycycline and Ciprofloxacin	Standard blood and body fluid precautions; noncontagious between humans	Wash hands and immediate area, hands with antimicrobial soap and surfaces with 5% bleach slolution
Brucellosis *Brucella suis*	Inhalation of bacteria; systemic infection through ingestion of contaminated food or contact with infected animals	Prevent contact; universal body fluid precautions are suggested	Decontaminate hands and area Doxycycline and Rifampin or Septra	If lesions are present use universal precautions prevent contact	Wash hands and immediate area, hands with antimicrobial soap and surfaces with 5% bleach slolution
Plague *Yersinia pestis*	Pneumonic plague highly contagious through aerosolized droplets	Prevent contact; this illness is tranmitted by large particle droplets expelled from the mouth and nose	Decontaminate hands and limit contact; Streptomycin or Chloramphenicol or Doxycycline	Cohort infected patients; care-giver requires protection in the form of a mask and gloves	Wash hands and immediate area, hands with antimicrobial soap and surfaces with 5% bleach slolution
Tularemia *Francisella tularensis*	Inhalation of aerosol; cutaneous infection through broken skin	Prevent contact with aerosol; noncontagious between humans	Decontaminate hands and limit contact; Streptomycin or Gentamicin	Routine decontamination of surfaces; noncontagious between humans	Wash hands and immediate area, hands with antimicrobial soap and surfaces with 5% bleach slolution

APPENDIX B

Biological Agent Quick Reference Guide: Viruses

Virus	Transmission	Precautions	Treatment	Isolation	Protective Measures
Equine Encephalitis	Mosquito	Insect repellent	Supportive	None	Limited exposure to insect environment
Viral Hemorrhagic Fevers	Aerosolized droplets and patient contact	Extremely contagious	Supportive	Strict isolation, contact precautions must be in place. Requires dedicated equipment and specialized patient handling	Limit patient contact, caregiver must have protective equipment, total decontamination of all supportive supplies and supplies used.
Small Pox (Variola virus)	Highly contagious between persons from airborne droplet exposure, and contact with lesions	Prevent contact Extremely contagious	Consult infectious disease physician	Strict isolation, contact precautions must be in place. Requires dedicated equipment and specialized patient handling	Limit patient contact, caregiver must have protective equipment, total decontamination of all supportive supplies and supplies used.

APPENDIX C

Biological Agent Quick Reference Guide: Toxins

Toxin	Transmission	Precautions	Treatment	Isolation	Protective Measures
Ricin	Ingestion, injection, or inhalation of toxin	Prevent contact with substance; noncontagious between humans	Supportive	Noncontagious between humans	Wash hands and immediate area, hands with antimicrobial soap and surfaces with 5% bleach slolution
T-2 Mycotoxins	Ingestion, injection, or inhalation of toxin	Prevent contact Prevent contact noncontagious between humans	Supportive	Noncontagious between humans	Wash hands and immediate area, hands with antimicrobial soap and surfaces with 5% bleach slolution
Botulism *Clostridium botulinium*	Ingestion of toxin; inhalation of aerosolized toxin	Prevent contact with the aerosolized toxin; noncontagious between humans	Decontaminate hands and immediate area; equine antitoxin	Routine decontamination of surfaces; Noncontagious between humans	Wash hands and immediate area, hands with antimicrobial soap and surfaces with 5% bleach slolution
Staphylocccal Enterotoxin B (SEB)	Ingestion of contaminated water or food source	Prevent contact with the toxin; noncontagious between humans	Decontaminate hands and immediate area	Routine decontamination of surfaces; Noncontagious between humans	Wash hands and immediate area, hands with antimicrobial soap and surfaces with 5% bleach slolution

APPENDIX D

Chemical Agent Quick Reference Guide

Chemical	Transmission	Precautions	Treatment	Isolation	Protective Measures
Nerve Agents *Organophosphates*	Liquid that has been aerosolized	Exclude from area; decontamination with soap and water if exposed	Specific medical antidote	Retreat to safe room	Maintain isolation from the external environment until notified by authorities or rescued by emergency service personnel
Vesicants *Acids or alkalis*	Liquid that has been aerosolized	Exclude from area; decontamination with soap and water if exposed	Supportive	Retreat to safe room	Maintain isolation from the external environment until notified by authorities service personnel
Blood Agents *Cyanides or arsine*	Gases or liquids that vaporize	Exclude from area; decontamination with soap and water if exposed	Extremely Lethal	Retreat to safe room; if incapacitated death occurs within seconds	Maintain isolation from the external environment until notified by authorities service personnel
Choking Agents *Chlorine, Phosgene*	Gases	Exclude from area; decontamination with soap and water if exposed	Supportive	Retreat to safe room	Maintain isolation from the external environment until notified by authorities service personnel
Riot Control *Tear gas, Mace, Pepper spray*	Aerosolized liquids or particulate solids	Exclude from area; decontamination with soap and water if exposed	Supportive the effects last for 20–30 minutes	Retreat to safe room	Maintain isolation from the external environment until notified by authorities service personnel

APPENDIX E

Emegency Preparedness Safety Checklists

Action	Details
Family Disaster Plan	
Establish communication options	Post emergency telephone numbers Ensure at least one non-radio phone in the house Secure a friend or relative as a contact Have at least two forms of contact with friend or relative: e-mail, cell phones, home phone, work phone, and/or pagers Have an out-of-state emergency contact
Stock food	Store nonperishable emergency food supplies Ensure you have a nonelectric can opener
Plan emergency escape routes and meeting places	Pre-establish your family's contact meeting point in case an evacuation is called; this can be done through your emergency contact
Establish safe room or area	Establish an improvised home shelter using plastic sheeting and tape to secure a safe environment within your home; shut off heating and ventilation; use plastic to cover windows and doors, enclosing a large area (kitchen, dining room, living room, or bathroom, for example) within your house or apartment
Safe Room should have an area where toilet facilities are separated	Make your safe room area large enough to incorporate a place where waste can be kept and personnel hygiene maintained Plan toilet facilities for your safe room

Emegency Preparedness Safety Checklists (continued)

Action	Details
Emergency Disaster Kit	
Water	Store two gallons per person per day, for 3–7 days
Food	Store canned nonperishable food supplies
Radio and batteries	Store 3–7 day supply of batteries
Plastic eating ware	Reduce water use and maintain clean environment
Flashlights and batteries	Store 3–7 day supply of batteries
Blankets, pillows, and clothing	Especially prepare for cold environments
Vehicles with full tank of gas	Have enough gas to lead you to safety, in case an evacuation is called
Cash	Have cash on hand since banks and ATMs may be out of service for an extended period of time
First-aid supplies	Have emergency first-aid supplies on hand to take care of minimal injuries since emergency services may not be able to get to you
Special needs items	Store additional supplies for special needs individuals in your plan; infants and the elderly may need supplies that are not recognized by most
Toiletries	Maintain personal hygiene (feminine hygiene and geriatric hygiene items)
Waste containers	Store plastic bags and containers for the disposal of waste

INDEX

A

C

D

E

N

O

P

W

Y

Z